Karl Kuhlemann
Nonstandard Analysis

Also of Interest

Philosophy of Mathematics
Thomas Bedürftig, Roman Murawski, 2018
ISBN 978-3-11-046830-4, e-ISBN (PDF) 978-3-11-046833-5,
e-ISBN (EPUB) 978-3-11-047077-2

Abstract Algebra. With Applications to Galois Theory, Algebraic Geometry, Representation Theory and Cryptography, 2024
Gerhard Rosenberger, Annika Schürenberg, Leonard Wienke, 2024
ISBN 978-3-11-113951-7, e-ISBN (PDF) 978-3-11-114252-4,
e-ISBN (EPUB) 978-3-11-114284-5

Mathematical Logic. An Introduction
Daniel Cunningham, 2023
ISBN 978-3-11-078201-1, e-ISBN (PDF) 978-3-11-078207-3,
e-ISBN (EPUB) 978-3-11-078219-6

Multivariable and Vector Calculus
Joseph D. Fehribach, 2024
ISBN 978-3-11-139238-7, e-ISBN (PDF) 978-3-11-139348-3,
e-ISBN (EPUB) 978-3-11-139428-2

Applied Nonlinear Functional Analysis. An Introduction
Nikolaos S. Papageorgiou, Patrick Winkert, 2024
ISBN 978-3-11-128421-7, e-ISBN (PDF) 978-3-11-128695-2,
e-ISBN (EPUB) 978-3-11-128832-1

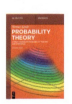
Probability Theory. A First Course in Probability Theory and Statistics
Werner Linde, 2024
ISBN 978-3-11-132484-5, e-ISBN (PDF) 978-3-11-132506-4,
e-ISBN (EPUB) 978-3-11-132517-0

Karl Kuhlemann

Nonstandard Analysis

In Higher Education, Logic and Philosophy

DE GRUYTER

Mathematics Subject Classification 2020
26E35, 97I10, 00A30

Author
Dr. Karl Kuhlemann
Altenberge
Germany
kus.kuhlemann@t-online.de

ISBN 978-3-11-142887-1
e-ISBN (PDF) 978-3-11-142968-7
e-ISBN (EPUB) 978-3-11-143053-9

Library of Congress Control Number: 2024944096

Bibliographic information published by the Deutsche Nationalbibliothek
The Deutsche Nationalbibliothek lists this publication in the Deutsche Nationalbibliografie;
detailed bibliographic data are available on the Internet at http://dnb.dnb.de.

© 2025 Walter de Gruyter GmbH, Berlin/Boston
Cover image: Eoneren / E+ / Getty Images
Typesetting: VTeX UAB, Lithuania
Printing and binding: CPI books GmbH, Leck

www.degruyter.com

With love for my wife Susanne

Preface

How can there be infinitesimal real numbers? How can there be infinitely large natural numbers? And how can such numbers be used to teach calculus and at the same time provide interesting insights into the foundations of mathematics? This book provides answers to these questions and explains the theoretical framework that makes them possible. It is thus part textbook, part monograph.

The textbook part (Chapters 1 and 2) maintains a level appropriate to undergraduate studies and is therefore suitable as a basis for a course supplementing the analysis lectures. The monograph part (Chapters 3 to 7) presents the theoretical background in more detail and discusses it in an extended context under the aspects of higher education, logic, and philosophy.

Chapter 1 begins with an introductory example, gives a preliminary answer to the question of why it might be worthwhile to deal with nonstandard analysis, and takes a critical look at the conventional teaching of analysis.

Currently, nonstandard analysis is hardly considered in university teaching. We argue that nonstandard analysis is not only valuable for teaching, but also for the understanding of standard analysis and mathematics as a whole. An axiomatic approach that takes into account different levels of language (for example, in distinguishing between sums of ones and the natural numbers of the theory) leads naturally to a nonstandard theory. Historical ideas from Leibniz can be used as motivation. Chapter 2 contains an elaborated concept that follows this path.

Chapter 3 introduces the most important approaches to nonstandard analysis. We discuss in detail the omega calculus of Schmieden and Laugwitz (which itself does not yet belong to nonstandard analysis, but is a good preparation to it), Robinson's model-theoretic approach and Nelson's internal set theory. Other axiomatic approaches related to internal set theory are outlined in less detail, focusing on the differences between them.

Chapter 4 shows how the different approaches can be used in basic analysis. We take a look at common analysis textbooks, report on a survey of analysis lecturers, and discuss common concerns about nonstandard analysis.

In Chapter 5, we explore how nonstandard analysis challenges familiar ideas about mathematics. In particular, we deal with the concepts of infinity, continuum, set, and natural number.

In Chapter 6, we discuss logical, model- and set-theoretical investigations in order to point out possible mathematical reasons that may lead to reservations about nonstandard analysis. Various foundational positions as well as ontological, epistemological, and application-related questions are also addressed.

Chapter 7 summarizes the key findings of the previous chapters and discusses possible consequences for teaching.

Overall, the aim of this book is to show how nonstandard analysis can usefully complement introductory courses and promote awareness of the foundations of calculus and mathematics in general. This could be particularly interesting for students and teachers of calculus, but also for users of calculus who are interested in the foundations of this subject.

I would like to thank Mikhail G. Katz (Bar-Ilan University, Ramat Gan) for his numerous helpful comments on the manuscript, as well as Markus Haase (Christian-Albrechts-Universität zu Kiel), Thomas Bedürftig (Leibniz Universität Hannover) and Roman Murawski (Adam Mickiewicz University, Poznań), who contributed with valuable advice to the present result. I am grateful to the de Gruyter publishing house for the opportunity to publish the book, especially to Nadja Schedensack and Kristin Berber-Nerlinger for their excellent support during the publication process.

Altenberge, October 2024   Karl Kuhlemann

Contents

Preface —— VII

List of Figures —— XV

List of Tables —— XVII

1 **Introduction** —— 1
1.1 An example to start with —— 1
1.2 Why should we concern ourselves with nonstandard analysis? —— 2
1.3 The real numbers in conventional analysis courses —— 6
1.3.1 Axiomatic introduction —— 6
1.3.2 The natural numbers —— 7
1.3.3 Naive numbers and theoretical numbers —— 8
1.3.4 Construction instead of axiomatics? —— 9
1.3.5 Mathematical induction and recursive definitions —— 9
1.3.6 Finite sets —— 10
1.3.7 The implicitly used set axioms —— 11
1.3.8 The question of mathematical rigor —— 12

2 **Elements of a complementary calculus course** —— 14
2.1 Leibniz as a historical point of reference —— 15
2.2 A modern translation of Leibniz's ideas —— 16
2.3 Preliminary remarks on axiomatics —— 17
2.4 Set theory in analysis —— 19
2.4.1 The standard set axioms —— 19
2.4.2 The set axioms not used —— 22
2.4.3 Cartesian products —— 23
2.4.4 Relations and functions —— 24
2.5 The Archimedean field of real numbers —— 24
2.5.1 The field axioms —— 25
2.5.2 The order axioms —— 25
2.5.3 Embedding the natural numbers —— 25
2.5.4 Definitions and proofs in the metalanguage —— 26
2.5.5 Again: Mathematical induction and recursive definitions —— 27
2.5.6 Again: Finite sets —— 30
2.5.7 The Archimedean axiom —— 31
2.5.8 Interim conclusion —— 31
2.6 The nonstandard axioms of analysis —— 32
2.6.1 Axiom schema of transfer —— 32
2.6.2 Idealization axiom for real numbers —— 34

2.6.3	The standard part axiom and the completeness of real numbers —— **36**	
2.7	Concluding remarks on the axiomatics —— **38**	
2.8	External criteria for central concepts of analysis —— **40**	
2.8.1	Convergence of sequences —— **41**	
2.8.2	Continuity at a point —— **43**	
2.8.3	Uniform continuity —— **44**	
2.8.4	Accumulation point of a set —— **44**	
2.8.5	Limits of functions —— **45**	
2.8.6	Differentiability —— **46**	
2.8.7	Integrability —— **47**	
2.9	Some theorems and proofs —— **49**	
2.9.1	Theorems about sequences —— **49**	
2.9.2	Theorems about continuous functions —— **49**	
2.9.3	Theorems of differential calculus —— **51**	
2.9.4	Theorems of integral calculus —— **53**	
2.9.5	Fundamental theorem of calculus —— **55**	
2.10	Nonstandard definitions —— **56**	
2.10.1	Axiom schema of standardization —— **56**	
2.10.2	Implicit definitions via standardization —— **58**	
2.11	Outlook —— **60**	
2.11.1	Formalization and conservativeness —— **60**	
2.11.2	Countable idealization —— **61**	
2.11.3	Parameter-free standardization —— **62**	
3	**An overview of nonstandard analysis —— 63**	
3.1	The history of nonstandard analysis —— **63**	
3.2	Omega calculus according to Schmieden and Laugwitz —— **66**	
3.2.1	Definition of omega numbers —— **67**	
3.2.2	Extension of relations and functions —— **68**	
3.2.3	Infinite and infinitesimal omega numbers —— **69**	
3.2.4	Hyperfinite sequences and sums —— **70**	
3.2.5	Central concepts of elementary analysis with omega numbers —— **71**	
3.2.6	Limitations —— **72**	
3.3	The hyperreal numbers —— **73**	
3.3.1	Field extension via ultrafilter —— **73**	
3.3.2	Transfer principle for hyperreal numbers —— **74**	
3.3.3	Field extension via compactness theorem —— **75**	
3.4	Nonstandard analysis in superstructures —— **76**	
3.4.1	Transitively bounded formulas —— **77**	
3.4.2	Superstructures —— **77**	
3.4.3	Elementary embeddings —— **78**	
3.4.4	Nonstandard embeddings —— **80**	

3.4.5	Standard elements and standard sets —— 80	
3.4.6	Internal elements and internal sets —— 81	
3.4.7	Hyperinfinite sets —— 82	
3.4.8	Analysis for internal functions —— 83	
3.4.9	Implicit definitions via $*$ —— 84	
3.4.10	Enlargements and saturation —— 85	
3.4.11	The existence of nonstandard embeddings —— 87	
3.5	Internal set theory —— 87	
3.5.1	Extension of the language —— 87	
3.5.2	Transfer —— 89	
3.5.3	Idealization —— 90	
3.5.4	Standardization —— 91	
3.5.5	Elementary analysis in internal set theory —— 92	
3.6	Other axiomatic approaches —— 93	
3.6.1	Bounded set theory —— 93	
3.6.2	External set theories —— 94	
3.6.3	Relative set theories —— 97	
4	**On practice and acceptance in teaching —— 101**	
4.1	Textbooks on analysis —— 101	
4.2	Nonstandard introductions to analysis —— 101	
4.2.1	Construction with Fréchet filter —— 101	
4.2.2	Generalized Ω-adjunction —— 104	
4.2.3	Superreal numbers —— 105	
4.2.4	Axiomatic introduction of hyperreal numbers —— 107	
4.2.5	Elementary calculus based on internal set theory —— 109	
4.2.6	Elementary calculus based on relative set theory —— 110	
4.2.7	Summary and comparison —— 112	
4.3	Teaching experience —— 113	
4.3.1	The cognitive advantage hypothesis —— 113	
4.3.2	The cognitive existence of infinitesimals —— 116	
4.3.3	A-track vs. B-track —— 118	
4.3.4	Is there resistance? —— 119	
4.4	A survey among analysis teachers —— 121	
4.4.1	Responses regarding current practice —— 121	
4.4.2	The assessment of nonstandard analysis (from the teachers' perspective) —— 122	
4.4.3	The teachers' arguments —— 122	
4.4.4	Reasons for a negative attitude towards nonstandard analysis in teaching —— 122	
4.4.5	How difficult is nonstandard analysis? —— 124	
4.4.6	How relevant is nonstandard analysis? —— 126	

4.4.7	What are the benefits of nonstandard analysis?	130
4.4.8	Are there other reasons for a negative attitude?	132

5 Philosophical questions and mathematical challenges — 133

5.1	From the foundations of mathematics	133
5.1.1	A look at history	133
5.1.2	Is there right and wrong mathematics?	134
5.1.3	Fundamental questions of the philosophy of mathematics	135
5.1.4	Realism	136
5.1.5	Constructivism	139
5.1.6	Formalism	140
5.1.7	Reverse mathematics and Hilbert's program	142
5.1.8	Mathematical practice	145
5.2	Infinity	146
5.2.1	Potential vs. actual infinity	146
5.2.2	Arithmetic vs. unarithmetic infinity	147
5.2.3	Challenge: The actual infinite in arithmetic	151
5.3	The continuum	152
5.3.1	The essence of the continuum	152
5.3.2	Continuum vs. set	153
5.3.3	The Cantor–Dedekind postulate	153
5.3.4	The Archimedean axiom	154
5.3.5	Challenge: The non-Archimedean continuum	155
5.4	Set theory	156
5.4.1	The relevance of set theory in mathematics	156
5.4.2	The set universe as a cumulative hierarchy	157
5.4.3	The constructible hierarchy	158
5.4.4	Variety of models	159
5.4.5	Metalanguage and object language	160
5.4.6	Background set theory and object set theory	160
5.4.7	The so-called standard models	161
5.4.8	Nonstandard pictures of set theory	163
5.4.9	Multiverse theories	164
5.4.10	The role of the axiom of choice	165
5.4.11	Challenge: Nonstandard set universes	167
5.5	The natural numbers	168
5.5.1	Peano structures	168
5.5.2	Object numbers and metanumbers	169
5.5.3	Which are the true natural numbers?	170
5.5.4	Challenge: Unlimited natural numbers	171

6 From the foundations of mathematics: On the status of nonstandard numbers — 173

- 6.1 Ontological questions — 173
- 6.1.1 Are nonstandard numbers contradictory? — 173
- 6.1.2 Do nonstandard numbers exist? — 175
- 6.1.3 Summary of the ontological answers — 179
- 6.2 Epistemological questions — 179
- 6.2.1 How definite is nonstandard analysis? — 179
- 6.2.2 How reliable is nonstandard analysis, foundationally speaking? — 184
- 6.2.3 Summary of the epistemological answers — 184
- 6.3 Questions on applicability — 184
- 6.3.1 What can nonstandard numbers mean for the real world? — 185
- 6.3.2 Is nonstandard analysis applicable to the real world? — 186
- 6.3.3 An example from probability theory — 188
- 6.3.4 Summary of the answers on applicability — 190

7 Conclusion — 192

- 7.1 Summary of the results — 192
- 7.1.1 Nonstandard analysis in higher education — 192
- 7.1.2 The teachers' assessment — 192
- 7.1.3 On the negative attitude towards nonstandard analysis — 192
- 7.1.4 Critical reflection on the reasons for rejection — 195
- 7.2 What is the bottom line? — 198
- 7.2.1 Options for teaching — 198
- 7.2.2 Closing words — 199

Bibliography — 201

Index — 211

List of Figures

Figure 2.1	Nonstandard proof of the least-upper-bound property.	**37**
Figure 2.2	The set of natural numbers in the nonstandard view.	**39**
Figure 2.3	Nonstandard proof of the fundamental theorem.	**56**
Figure 4.1	Number of published documents with MSC 26E35.	**128**
Figure 4.2	Number of published documents with MSC 03H05.	**128**
Figure 4.3	Number of published documents with MSC 03H10.	**128**
Figure 4.4	Number of published documents with MSC 03H15.	**129**
Figure 4.5	Number of published documents with MSC 28E05.	**129**
Figure 4.6	Number of published documents with MSC 54J05.	**129**

List of Tables

Table 4.1	Summary and comparison of elementary nonstandard introductions to calculus.	—— 113
Table 4.2	Number of Students Attempting a Solution ([210]).	—— 114
Table 4.3	Student Responses to Question 3 ([210]).	—— 115
Table 4.4	Instructor Questionnaire: Part One ([210]).	—— 115
Table 4.5	Instructor Questionnaire: Part Two ([210]).	—— 115
Table 4.6	Result of the survey at the beginning of the course ([216], p. 5).	—— 117
Table 4.7	Result of the survey at the end of the course ([216], p. 5).	—— 117
Table 4.8	Result of the poll by Katz and Polev ([114]).	—— 119
Table 4.9	Reasons given by teachers for (+) or against (−) the use of nonstandard analysis in teaching.	—— 123

1 Introduction

1.1 An example to start with

A simple task from Calculus 1 is: Determine the limit of the infinite geometric series

$$q^0 + q^1 + q^2 + q^3 + \cdots,$$

where q is any real number with $|q| < 1$.

The first important step towards a solution is to recognize that the following formula holds for the partial sums of the series:

$$q^0 + q^1 + q^2 + q^3 + \cdots + q^n = \frac{1 - q^{n+1}}{1 - q}. \tag{1.1}$$

The classical proof then continues by showing that (q^{n+1}) is a null sequence using previously proven theorems about limits of sequences (i.e., ultimately using an ε-argument) and $(\frac{1-q^{n+1}}{1-q})$ therefore has the limit $\frac{1}{1-q}$.

What do you think of the following proof? In (1.1), we sum up to an infinite index n and ignore any infinitely small parts in the result. Since with n also $n + 1$ is infinite and therefore q^{n+1} is infinitely small, this also leads to the limit $\frac{1}{1-q}$. This would be a proof in the style of Johann Bernoulli or Leonhard Euler. However, such a proof would hardly be accepted in an analysis exam today, as it raises a number of questions:

1. What are infinite and infinitely small numbers in this context?
2. What should q^k mean for infinite k, an infinite multiplication of q by itself?
3. What does summing up to an infinite index n mean?
4. Does the formula (1.1) also apply to infinite n?

We give the following provisional answer to the first question: A number can be considered "infinite" if it is greater than any real number that can be defined in ordinary analysis (e.g., by a term or a formula), i.e., in particular greater than any sum of ones like $1, 1 + 1, 1 + 1 + 1$, and so on. To avoid confusion with the set-theoretic term *infinite*, we will prefer to call such numbers *unlimited*. Later, in Section 2.6.2, we will give a rigorous definition of the term. *Infinitely small* or *infinitesimal* numbers are reciprocals of unlimited numbers.

Let us put the other questions on the back burner for a moment and ask instead—for the sake of fairness—a question about the classic proof: From which set may we take the values for n in (1.1)? The answer seems clear: from the set $\mathbb{N}_0 := \mathbb{N} \cup \{0\}$, but how is the set $\mathbb{N}$ defined in calculus? The "naive definition" $\mathbb{N} := \{1, 2, 3, \dots\}$ is ruled out because it is not a definition. What is possible and can be found in numerous textbooks is the definition of $\mathbb{N}$ as the intersection of all inductive subsets of $\mathbb{R}$ (see Section 1.3.2).

However, this definition by no means excludes the existence of unlimited natural numbers (in the sense of our provisional definition). The deeper reason for this initially surprising statement is that one has to distinguish the naive natural numbers (with which, for example, one counts the number of ones in a sum term) from the natural numbers of a mathematical theory such as analysis or set theory (see Section 1.3.3).

But if the existence of unlimited numbers in $\mathbb{N}_0$ cannot be ruled out, then questions 2 to 4 above must also be asked analogously for classical analysis: What should q^k mean for general (possibly unlimited) $k \in \mathbb{N}_0$? What should summation up to an arbitrary (possibly unlimited) index $n \in \mathbb{N}_0$ mean? Does the formula (1.1) apply to any (possibly unlimited) $n \in \mathbb{N}_0$?

Of course, classical analysis has answers to this because, strictly speaking, the powers and partial sums in (1.1) are not defined by product terms $q \cdots q$ or sum terms $q^0 + \cdots + q^n$, but are defined recursively, i.e., by $q^0 := 1$ and $q^{k+1} := q \cdot q^k$ or $s_0 := q^0$ and $s_{n+1} := s_n + q^{n+1}$. Therefore, they are defined for *all* $n \in \mathbb{N}_0$ (according to the recursion theorem). And the formula (1.1) is proven by mathematical induction and therefore applies to *all* $n \in \mathbb{N}_0$ (according to the theorem of mathematical induction).

It is just that: The recursion theorem and the theorem of mathematical induction apply regardless of whether there are unlimited numbers in $\mathbb{N}_0$ or not. They simply follow from the definition of $\mathbb{N}_0$ and some set theory (which, however, is generally used intuitively in analysis courses without explicitly stating axioms). Questions 2 to 4 therefore raise bogus problems in relation to unlimited numbers.

In order to justify the proof based on Bernoulli and Euler reasoning, the question remains whether there are unlimited numbers in $\mathbb{N}_0$. With an axiomatic approach, this is a question of the assumptions made, and making assumptions that lead to unlimited numbers in $\mathbb{N}_0$ is demonstrably possible without additional risk and is no more daring than the assumption that infinite sets exist, which is generally accepted today with the axiom of infinity in ZFC (the Zermelo–Fraenkel axiom system with the axiom of choice). This results from the investigation of nonstandard theories such as Edward Nelson's internal set theory and related theories that are conservative extensions of ZFC.

In Chapter 2, we show how a simplified version of such a nonstandard theory can be used for basic analysis. Before that, we ask why we should deal with nonstandard analysis at all (Section 1.2), and we take a closer look at the real numbers as they are introduced in conventional analysis courses (Section 1.3).

1.2 Why should we concern ourselves with nonstandard analysis?

Today's analysis emerged from the infinitesimal calculus of Leibniz and Newton. For two centuries, infinitesimals, i.e., infinitely small quantities, were part of the mathematicians' toolkit and gave the newly created discipline an extraordinary boost before they seemed to be dispensable due to Weierstrass' limit concept and the construction of the real numbers by Cantor and Dedekind. Finally, infinitesimals were banished from analysis.

Almost a century later, they returned in the context of *nonstandard analysis*—now strictly based on model theory. But why should we deal with infinitesimals again if we had successfully got rid of them before and analysis could obviously do without them?

Detlef Laugwitz answered the question "What is infinitesimal mathematics and what is it used for?" in 1978 as follows:

> So why do we do any mathematics? I see three possible justifications in particular,
> 1. that of application in the broadest sense, be it in the solution of problems within and outside mathematics, be it through the new or further development of methods;
> 2. that of teaching, either by contributing to the content or by improving the way it is taught;
> 3. that of reflection on mathematics itself, be it its history or its further development.
>
> ([129], p. 10, own translation).[1]

In principle, the justifications given can still claim validity today. Nonstandard analysis has enriched the tools of mathematics and produced results in various mathematical and mathematically oriented fields, for example, in topology, stochastics, functional analysis, but also in mathematical physics or economics (see Section 4.4.6). It can make proofs simpler and clearer, support the teaching of basic concepts of analysis and contribute to the understanding of the historical infinitesimal calculus. The value of nonstandard analysis is therefore measured in various dimensions: mathematical, didactic, historical.

The historical perspective in particular is quite unusual. Today, the heyday of infinitesimal calculus is often seen as a time when gifted mathematicians, using questionable methods, arrived at profound and correct results that can now finally be rigorously justified within the framework of modern analysis, be it the fundamental theorem, the mathematical description of the "catenary", Leibniz's series formula for π or Euler's famous equation $e^{i\pi} + 1 = 0$. However, Laugwitz considered it wrong to see the reference to history only in this way. Rather, by studying the old writings, he saw an opportunity to uncover sources that had been partially buried by conventional analysis and that could now have a stimulating effect again through nonstandard analysis (cf. [129], pp. 13–14). He himself gave numerous examples of this, from Leibniz and Bernoulli to Euler and Cauchy.[2]

---

[1] In the original German: "Wozu also betreibt man irgendeine Mathematik? Ich sehe vor allem drei mögliche Rechtfertigungsgründe,
1. den der Anwendung im weitesten Sinne, sei es in der Lösung von Problemen innerhalb und außerhalb der Mathematik, sei es durch die Neu- oder Weiterentwicklung von Methoden;
2. den des Unterrichts, sei es durch Beiträge zu den Inhalten oder durch eine Verbesserung der Vermittlung;
3. den der Reflexion auf die Mathematik selbst, seien es ihre Geschichte oder ihre Weiterentwicklung."

[2] The role of infinitesimals, especially in the works of Leibniz and Cauchy, is still controversial among historians today. See, for example, [177, 13, 117, 6, 14, 15].

Mathematics—like any other science—is subject to historical change. Its methods and concepts evolve. This also applies to analysis. While this statement seems self-evident in relation to the past, it seems almost disconcerting with regard to the future. From our current perspective, calculus—at least as far as its foundations are concerned—appears to have come to a satisfactory conclusion with Weierstrass's limit concept and the set-theoretical construction of real numbers, and all the struggle for today's established concepts seems like a consistent striving towards the achieved goal. But today's calculus must be seen neither as a goal nor as the conclusion of an inevitable development. Laugwitz once again:

> For me, broadening our understanding of historical development is one of the main motives for working with infinitesimal methods. We are still a long way from a systematic and reasonably complete historiography. Many accounts are based on a finalist view: Development is seen from the present state, with this appearing as the goal of history. Now that we know several theories for the justification of analysis, the historiography—even the finalistically oriented one—must be revised. The same applies to the relationship between numbers and the continuum, now that the identification of the continuum with $\mathbb{R}$ has been called into question again ([130], p. 254, own translation).[3]

A current critique of a teleological view of the analysis development can be found, for example, in [14], Section 4.

Whether history could have taken a different course remains a matter of speculation. The actual development has led to what we now call conventional or "standard" analysis. What role does nonstandard analysis play in this situation? Is it merely a logical gimmick, an unnecessary undertaking to provide a serious foundation for outdated and superfluous ideas from a dark era? Or is it rather the analysis of the future? In the foreword to the second edition of Robinson's book *Non-standard Analysis*, Kurt Gödel took a clear position on this:

> I would like to point out a fact that was not explicitly mentioned by Professor Robinson, but seems quite important to me; namely that nonstandard analysis frequently simplifies substantially the proofs, not only of elementary theorems, but also of deep results. This is true, e. g., also for the proof of the existence of invariant subspaces for compact operators, disregarding the improvement of the result; and it is true in an even higher degree in other cases. This state of affairs should prevent a rather common misinterpretation of nonstandard analysis, namely the idea that it is some kind of extravagance or fad of mathematical logicians. Nothing could be farther from the truth. Rather

---

[3] In the original German: "Für mich ist die Erweiterung des Verständnisses für die historische Entwicklung ein Hauptmotiv für die Beschäftigung mit Infinitesimalmethoden. Von einer systematischen und einigermaßen vollständigen Geschichtsschreibung sind wir noch weit entfernt. Viele Darstellungen gehen von einer finalistischen Auffassung aus: Die Entwicklung wird vom gegenwärtigen Stand aus gesehen, wobei dieser als Ziel der Geschichte erscheint. Nachdem wir jetzt mehrere Theorien zur Begründung der Analysis kennen, ist die Geschichtsschreibung – sogar die finalistisch orientierte – zu revidieren. Analoges gilt für die Beziehungen zwischen Zahlen und Kontinuum, nachdem die Identifikation des Kontinuums mit $\mathbb{R}$ wieder in Frage gestellt ist."

there are good reasons to believe that nonstandard analysis, in some version or other, will be the *analysis of the future*.

One reason is the just mentioned simplification of proofs, since simplification facilitates discovery. Another, even more convincing reason, is the following: Arithmetic starts with the integers and proceeds by successively enlarging the number system by rational and negative numbers, irrational numbers, etc. But the next quite natural step after the reals, namely the introduction of infinitesimals, has simply been omitted. I think, in coming centuries it will be considered a great oddity in the history of mathematics that the first exact theory of infinitesimals was developed 300 years after invention of the differential calculus ([183], p. xvi, emphasis added).

Given the clear dominance of standard analysis, even half a century after this quote, Gödel's assessment does not seem to have come true. On the other hand, half a century is not very much in relation to the history of analysis as a whole. Even today's basic concepts of standard analysis, such as real number, function or set, have only become established within several decades. And Gödel certainly thought in longer periods of time, as he speculated about the retrospective view "in coming centuries".

"What does nonstandard analysis offer to our understanding of mathematics?" asks Robert Goldblatt in the preface to his 1998 "Lectures on the Hyperreals" and lists five features that an answer should include:

1. New definitions of familiar concepts, often simpler and more intuitively natural [...]
2. New and insightful (often simpler) proofs of familiar theorems [...]
3. New and insightful constructions of familiar objects [...]
4. New objects of mathematical interest [...]
5. Powerful new properties an principles of reasoning [...]

([85], pp. vii–viii).

If Laugwitz, Gödel, and Goldblatt are right, why is nonstandard analysis so hesitantly accepted in teaching?

In this book, the question of the value of nonstandard analysis will be raised again, focusing on aspects of education and foundations of mathematics. What approaches to nonstandard analysis are there, and to what extent are they suitable for an introduction to analysis? What do analysis teachers think about the use of nonstandard analysis in teaching? What effects do nonstandard theories have on the self-image of mathematics? The axiomatic approaches and recent model-theoretical results in particular give reason to critically question familiar ideas about standard and nonstandard mathematics. We will also examine the question of whether a preference for standard theory over nonstandard theories can be derived from certain philosophical positions on the foundations of mathematics.

But let us start with a look at the theory of real numbers as it is usually taught in analysis courses.

1.3 The real numbers in conventional analysis courses

The real numbers can be constructed from the rational numbers (using set-theoretical means) in various ways, for example, by means of Cauchy sequences or Dedekind cuts. This work was essentially carried out by Cantor and Dedekind in 1872 [45, 59]. This year can therefore be considered the birth year of the real numbers. Another construction method uses sequences of nested intervals [12]. Axiomatic descriptions of the real numbers go back to Hilbert [96] and Tarski [221].

Overall, the number system can be constructed in the pure set theory ZFC (even already in Z^0)[4] via the intermediate stages $\mathbb{N}_0$, $\mathbb{Z}$, $\mathbb{Q}$, $\mathbb{R}$, $\mathbb{C}$, where $\mathbb{N}_0$ is identified with the lowest transfinite ordinal number ω (see Section 5.2.2). All complete ordered fields are uniquely isomorphic to $(\mathbb{R}, 0^\mathbb{R}, 1^\mathbb{R}, +^\mathbb{R}, \cdot^\mathbb{R}, <^\mathbb{R})$ (see [209], p. 591).

1.3.1 Axiomatic introduction

In analysis courses, the real numbers are usually introduced axiomatically. However, an axiom system that characterizes the real numbers requires second-order logic, i.e., a logic that can quantify not only over individuals (i.e., real numbers), but also over relations (e.g., properties of real numbers). This is due to the completeness axiom, which is essential for the characterization of the real number system. In a common semantic formulation (i.e., related to models) it reads: Every nonempty set of real numbers has a least upper bound.[5]

In analysis, however, higher-order sets are also considered (for example, sets of functions or sets of subsets of $\mathbb{R}$). Axiomatizations involving set theory are therefore common. Basically, these are extensions of an axiomatic set theory, whereby the set-theoretical axioms are usually not explicitly mentioned in analysis textbooks.

Royden and Fitzpatrick choose such an axiomatic approach in their textbook "Real Analysis".

> We assume that the set of real numbers, which is denoted by **R**, satisfies three types of axioms. We state these axioms and derive from them properties on the natural numbers, rational numbers, and countable sets ([187], p. 7).

The three types of axioms are the field axioms, the order axioms (in [187] the positivity axioms are used instead), and the completeness axiom. While the field axioms and the

---

[4] Here Z^0 contains the axioms of extensionality, pairing, union, power set, infinity, and the axiom schema of separation; Z also contains the axiom of regularity, ZF additionally contains the axiom schema of replacement (the axiom of pairing and the axiom schema of separation are then superfluous). In ZFC, the axiom of choice is added. For the formulation of the above axioms, see Section 2.4.

[5] First-order axiomatizations, where the completeness axiom is replaced by an axiom schema, always allow nonstandard models (see, for example, [26], pp. 321–322).

order or positivity axioms can still be formulated without any set vocabulary, this no longer applies to the completeness axiom, as we have seen above. We note, however, that the assumption of the existence of a set containing all real numbers is already an (unspoken) axiom that uses set vocabulary.

1.3.2 The natural numbers

In Section 1.1 we declared the naive definition $\mathbb{N} := \{1, 2, 3, \ldots\}$ inadmissible and claimed that it is not a definition. But what exactly is objectionable about this naive definition, which can certainly be found in introductions to calculus (see, for example, [1], p. 2)? It is the dots, of course.

The problem with the dots is not that we do not know how the list of elements continues. We do know that: After each number n comes its successor $n + 1$. The problem is rather that we do not know how it ends and what happens up to the closing brace. The naive answer should be: It *never* ends because after every number there is another one. Naively, we cannot set a closing brace at all. How do we then arrive at the total set of all naturals?

In an axiomatic theory, we only have the axioms, the primitive notions used in them and logic at our disposal to define new concepts. As it turns out, we must again include set theory in order to define the natural numbers within real analysis.

The axiomatic program of gaining the entire analysis from the axioms of the real numbers plus some set theory has the strange consequence that we initially do not know what the natural numbers are in analysis—at least not what they are in their *totality*. What we have are the numbers that result from the repeated addition of 1, i. e.,

$$1,$$
$$2 := 1 + 1,$$
$$3 := 1 + 1 + 1,$$

and so on. In this way, we obtain *potentially* infinitely many more natural numbers. What we do not yet have is the *set* $\mathbb{N}$ of *all* natural numbers. However, we need this set to define sequences and limits or to formulate the Archimedean property of $\mathbb{R}$.

Royden and Fitzpatrick show how this challenge is usually met:

> It is tempting to define the natural numbers to be the numbers $1, 2, 3, \ldots$, and so on. However, it is necessary to be more precise. A convenient way to do this is to first introduce the concept of an *inductive set* ([187], p. 11).

A subset of $\mathbb{R}$ is called *inductive* if it contains 1 and with each number x also its successor $x + 1$. Since the intersection of inductive sets is inductive again, the set of natural numbers can then be defined as the intersection of *all* inductive subsets of $\mathbb{R}$, i. e.,

$$\mathbb{N} := \bigcap \{E \in \mathcal{P}(\mathbb{R}) \mid E \text{ is inductive}\} \tag{1.2}$$

and

$$x \text{ is a natural number} \quad :\Leftrightarrow \quad x \in \mathbb{N}. \tag{1.3}$$

Here $\mathcal{P}(\mathbb{R})$ is the *power set* of $\mathbb{R}$, i.e., the set of all subsets of $\mathbb{R}$. Thus, $\mathbb{N}$ is the *smallest* inductive subset of $\mathbb{R}$ (in the sense of set inclusion). In other words, every inductive subset of $\mathbb{R}$ includes $\mathbb{N}$.

The above definition of $\mathbb{N}$ in (1.2) is possible because the axioms of set theory allow us
- to form the power set of a given set,
- to separate from a given set the subset of elements with a certain property (which can be formulated in the language of analysis),
- to form the intersection of a family of sets (i.e., a set whose elements are sets).[6]

If one has the set $\mathbb{N}$ available, one can define sequences and limits as usual, formulate and prove the Archimedean property of $\mathbb{R}$ and the convergence of all Cauchy sequences.

It is also possible to demand the Archimedean property of $\mathbb{R}$ and the convergence of all Cauchy sequences as axioms instead of the least-upper-bound property and to derive the latter as a theorem.

1.3.3 Naive numbers and theoretical numbers

Concepts without an axiomatic basis are often referred to as *naive* in mathematics, for example, when it comes to the concepts of number or set. Axioms and definitions based on them specify how concepts are to be used within a theory. And they decouple the concepts from the perception, intuition or everyday experience on which the naive use is based. This is also the case with natural numbers.

The natural numbers of an axiomatic theory are to be distinguished from the *naive* natural numbers of our everyday language. While the naive natural numbers are only *potentially* infinite (each number has a successor), the theoretical natural numbers (as a consequence of the axioms) are available in their totality (as elements of the *actual* infinite set $\mathbb{N}$).

In the terms of mathematical logic, the naive natural numbers belong to the *metalanguage*, i.e., the language in which we speak *about* the language of analysis (which in this context is called *object language*, see Section 5.4.5). We use the numbers of the metalanguage to specify, for example, how many free variables there are in a logical formula or how many summands there are in a summation term.

---

6 This actually already follows from the previous point (see Section 2.4.1).

For every naive natural number, we obtain a theoretical natural number (defined by a summation term with a corresponding number of ones). Conversely, however, we cannot formulate in the object language that every theoretical natural number can be represented by a sum of ones because the notion of a *sum term* is not available in the object language. It belongs to the metalanguage and denotes a certain construct of the object language, a certain type of string. For the same reason, we cannot define the set $\mathbb{N}$ in the object language as the set of all real numbers that can be represented as a sum of ones.

On the other hand, the necessary distinction between the natural numbers of different language levels offers the possibility of assuming the existence of unlimited natural numbers in theory, as Bernoulli and Euler did quite intuitively.

1.3.4 Construction instead of axiomatics?

In some textbooks, the real numbers are not introduced axiomatically, but are constructed. Rudin formulates the following theorem in Chapter 1 of his book "Principles of Mathematical Analysis" and proves it in the appendix:

Theorem ([188], p. 8). *There exists an ordered field $\mathbb{R}$ which has the least-upper-bound property. Moreover, $\mathbb{R}$ contains $\mathbb{Q}$ as a subfield.*

In "Foundations of Analysis", Landau starts even further back, with the natural numbers, and constructs the rational, real, and complex numbers on this basis:

> We assume the following to be given: A set (i.e. totality) of objects called natural numbers, posessing the properties—called axioms—to be listed below ([126], p. 1).

Then the (set-theoretically formulated) Peano axioms are listed, which uniquely characterize the natural numbers (see Section 5.5.1).

Does this rule out the possibility of unlimited natural numbers? Not at all. Regardless of whether one begins analysis directly with the real numbers or with the rational or natural numbers or whether one starts from a pure set theory, it is always necessary to distinguish the naive natural numbers of the metalanguage from the theoretical natural numbers, and there is always the option of assuming the existence of unlimited natural numbers.

1.3.5 Mathematical induction and recursive definitions

Due to the definition of $\mathbb{N}$, every inductive subset of $\mathbb{N}$ is equal to $\mathbb{N}$. This enables the following proof principle of *mathematical induction*: To show that a property $P(n)$ (which can be formulated in the object language) applies to all natural numbers, it must be shown that it applies to 1 and that for all $n \in \mathbb{N}$, $P(n)$ implies $P(n + 1)$. To justify this principle, consider the set $\{n \in \mathbb{N} \mid P(n)\}$ which, if it is inductive, must be equal to $\mathbb{N}$.

Mathematical induction used to prove statements about the theoretical natural numbers must be distinguished from the induction used in the metalanguage. The latter is employed when statements about inductively defined constructs of the object language (e. g., terms or formulas) are to be proven. One example is the generalization of the distributive law to summation terms of any length:

$$a \cdot (b_1 + \cdots + b_n) = a \cdot b_1 + \cdots + a \cdot b_n.$$

Here n is a natural number of the metalanguage that indicates the number of summands in a summation term. In contrast to the principle of mathematical induction, the principle of induction in the metalanguage is not a consequence of the axioms of the theory (for the justification of this principle, see Section 2.5.5).

An example of an inadmissible conflation of natural numbers of different language levels is the definition of the sequence of partial sums $(s_n)_{n \in \mathbb{N}}$ for a given sequence $(a_i)_{i \in \mathbb{N}}$ by

$$s_n := \sum_{i=1}^{n} a_i := a_1 + \cdots + a_n$$

for all $n \in \mathbb{N}$ (see, e. g., [188], p. 59).

The right-hand side $a_1 + \cdots + a_n$ is the abbreviation for a term of the object language (a summation term) that only makes sense for a natural number n of the metalanguage. This is not sufficient to define the sequence of partial sums. The recursive definition is correct:

$$s_1 := a_1, \quad s_{n+1} := s_n + a_{n+1}.$$

But here, too, there is a gap in the argumentation. The fact that recursive definitions are possible at all follows from the recursion theorem (which is usually not proven in analysis courses). The recursion theorem is based on the theorem of mathematical induction, which in turn follows from the definition of $\mathbb{N}$ as the smallest inductive subset of $\mathbb{R}$ (see Section 2.5.5).

1.3.6 Finite sets

Although most analysis textbooks define the terms *countable* and *uncountable* (and use Cantor's argument to show that $\mathbb{R}$ is uncountable), an explicit definition of the term *finite* is less common. However, this concept is less trivial than it first appears. Here, too, it is important to distinguish between the language levels.

Royden and Fitzpatrick give the following definition:

> A set E said to be *finite* provided either it is empty or there is a natural number for which E is equipotent to $\{1, \ldots, n\}$ ([187], p. 13).

Here $\{1, \ldots, n\}$ is the abbreviation of $\{k \in \mathbb{N} \mid 1 \leq k \leq n\}$. The empty set is included in this definition if $n = 0$ is allowed.

In everyday life, we only deal with finite sets. We are very familiar with dealing with them. The following exemplary statements therefore seem almost self-evident to us:
- If E is a finite set, then there is no bijection (one-to-one correspondence) between E and a proper subset of E.
- Every subset of a finite set is finite.
- The union of two finite sets is finite.
- The power set of a finite set is finite.
- Every finite nonempty subset of $\mathbb{R}$ contains a smallest number (minimum) and a largest number (maximum).

However, since the definition of finite sets is linked to the natural numbers, we must distinguish between the naive concept of finiteness in our everyday experience and the theoretical concept of finiteness in set theory. Therefore, in an axiomatic theory, even seemingly obvious statements about finite sets require proof. This is immediately obvious if you realize that there could be unlimited natural numbers.

The above statements can be proven by induction on the number of elements of the sets involved. First, however, it must be shown by induction that the size of a finite set (i. e., the number of its elements) is well defined at all. We will come back to this in Section 2.5.6.

1.3.7 The implicitly used set axioms

Since analysis makes use of set vocabulary, many textbooks or lecture notes contain a short introduction to (naive) set theory (see, for example, [18, 150, 187]). Care is always taken to ensure that set theory does not appear to belong to analysis. Rather, set theory appears to be a self-evident substructure that has nothing to do with the axiomatic theory. However, this separation is not justified from a logical point of view.

It is impossible to do axiomatic analysis with the vocabulary of set theory without axioms that rule the handling of sets. Why should we have to axiomatically demand the commutativity of the addition of real numbers, but not the possibility of forming union or power sets?

At least the following axioms from ZFC are implicitly used in an axiomatic approach to analysis (see Section 2.4.1):
- the axiom of extensionality,
- the axiom schema of separation,
- the axiom of pairing,

- the axiom of union sets,
- the axiom of power sets.

We also need an axiom that ensures the existence of a set containing all real numbers. The axiom of infinity from ZFC can then be dispensed with. For some proofs, the axiom of choice is also required. The axiom schema of replacement and the axiom of regularity are not usually used in analysis (see Section 2.4.2).

1.3.8 The question of mathematical rigor

In this Section 1.3 we have found that the distinction between language levels (related to the natural numbers, the concept of finiteness, recursive definitions and inductive proofs) is not always sufficiently transparent in conventional introductions to analysis and that the set theory used is not included in the axiomatics.

The teaching of analysis seems to be little affected by this at first. Set theory is used naively, but in the sense of the (not explicitly stated) axioms of ZFC. And analysis apparently works without a clean separation of language levels. It works because we work axiomatically when proving theorems and use the theoretical natural numbers and the theoretical "finite", although many people are more likely to think of the natural numbers and the "finite" of the metalanguage. Who feels the need to prove that subsets of a finite set are finite again?

The conflation of language levels and the omission of the axioms of set theory are a didactic compromise that deliberately leaves gaps in the application of the axiomatic method and makes concessions to the otherwise highly valued mathematical rigor. However, in this way we deprive ourselves of the possibility of allowing "arithmetically infinite", i.e., *unlimited* numbers in real analysis. One (perhaps intended) consequence of this approach is that the so-called standard theory appears (unjustifiably) as quite natural, while equally possible nonstandard theories are ignored.

If one distinguishes between the language levels from the outset and includes the required set theory in the axiomatics, this leads quite naturally to a potentially richer theory. It opens up the possibility of a nonstandard analysis within the real numbers. There is no need for a field extension of $\mathbb{R}$, no need for hyperreal numbers, because unlimited and infinitesimal numbers already slumber unrecognized in $\mathbb{R}$. At the same time, there are interesting starting points for a discussion of historical references, mathematical foundations and the self-image of mathematics. Such an approach therefore seems to be worth considering at least for courses accompanying or supplementing lectures on analysis.

In the basic lectures, an unbiased view towards nonstandard methods would be favored by paying attention to the following points:
- A (cautious) sensitization for the distinction of language levels and the distinction between naive and theoretical natural numbers.

- Preserving the option of unlimited numbers through precise formulation: Every sum of ones is a natural number, the reverse statement remains open.
- A (cautious) sensitization to the use of set-theoretical axioms (which, if not explicitly formulated as axioms, could at least be perceived as conscious additional assumptions).
- Introduction and experiments with limits *and* nonstandard methods in beginner lectures without any formal effort.

The inclusion of "infinitesimal" considerations alongside limit formalism serves not only to appreciate the historical roots of calculus, but also to understand mathematical foundations and to intuitively grasp conceptualizations and ideas of proof. As the further investigation will show, this is possible without danger, because unlimited and infinitesimal numbers are potentially available.

It is understandable that one does not want to spend much time on natural numbers, the distinction between language levels or set-theoretical considerations in basic calculus lectures. After all, the aim there is to get to the *proper* theorems of analysis as quickly as possible. In an elective course, however, there is the opportunity to devote a little more attention to these topics and thus learn something essential about axiomatic theories and thus about mathematics as a whole. In the following Chapter 2, we describe in more detail what such a course could look like.

2 Elements of a complementary calculus course

The introduction presented here is intended as an optional course that accompanies or supplements the basic lecture. We therefore assume that the basic concepts of calculus with their conventional definitions are known. The main objectives of the course are:
1. to raise awareness of the gaps in mathematical rigor in the basic lectures, of the set-theoretic background, and of the historical roots of calculus,
2. to stimulate reflection on the foundations of mathematics as an axiomatic-deductive science,
3. to present nonstandard analysis as an optional and effective enrichment of the spectrum of methods,
4. to offer an alternative and intuitive description of the basic concepts of analysis,
5. to establish the relationship between standard (i. e., conventional) analysis and nonstandard analysis,
6. to repeat and consolidate the material of the basic lecture from a new perspective.

We maintain a level appropriate for undergraduate studies and largely refrain from formalizing the language. Logical formulas and equivalence transformations only occur more frequently in the proofs of the equivalence of conventional definitions and the nonstandard criteria in Section 2.8. However, the explanations provided should make the transformations easy to understand.

Although the set theory used is included in the axiomatics, it is limited to the bare essentials. In particular, we do not need sophisticated concepts such as ultrafilters. The nonstandard axioms are a weakened version of the additional axioms from Edward Nelson's *internal set theory* [160]. We refer to the axiom system SPOT from [104], but add the axioms of the real numbers (as usual in introductory lectures on analysis) in order not to have to construct the real numbers in ZF.

We pay special attention to the distinction between the naive everyday numbers $1, 2, 3, \ldots$ and the natural numbers of the theory because this is a key to understanding axiomatic theories in general and to the optional enrichment with nonstandard numbers in particular.

Due to the objectives pursued, the path taken here is not the shortest and most direct way to incorporate infinitesimals into the theory of calculus. Such a way can be found, for example, in the axiomatic introduction of hyperreal numbers discussed in Section 4.2.4. With the elementary extension principle, nonstandard methods are immediately available, but at the price that the awareness of the foundations of mathematics and the relationship between standard and nonstandard mathematics remains in the background. Another way to teach calculus with infinitesimals is the differential-based approach, where differentials are treated informally as quantities rather than formally defined as hyperreal numbers (see, e. g., [72]).

2.1 Leibniz as a historical point of reference

Leibniz distinguished *assignable quantities* such as $1, \frac{2}{3}, \sqrt{2}, 10^{10^{10}}$ and *unassignable quantities* (quantities that we cannot assign because they are larger or smaller than any assignable quantity). Quantities were always positive for Leibniz.

In this historical terminology, a quantity was *infinite* if it was larger than any assignable quantity (in particular, larger than any assignable natural number $1, 2, 3, \ldots$), and *infinitely small* or *infinitesimal* if it was smaller than any assignable quantity (in particular, smaller than any assignable unit fraction $\frac{1}{2}, \frac{1}{3}, \frac{1}{4}, \ldots$).

Unassignable quantities should not differ in any way from assignable quantities—apart from the fact that they are not assignable. In a letter to his patron Varignon dated February 2, 1702, Leibniz described this principle as follows:

> ... et il se trouve que les regles du fini reussissent dans l'infini ... et que *viceversa*, les regles de l'infini reussissent dans le fini ... ([136], p. 26, original spelling preserved, emphasis in the original).[1]

The conclusion is that we can calculate with infinite and infinitesimal quantities as usual. On the question of their existence, Leibniz said:

> Nor does it matter whether there are such quantities in the nature of things, for it suffices that they be introduced by a fiction ...([138], p. 128, as translated by Arthur in [9], p. 27).[2]

Using such fictions, we can determine the slope of a smooth curve given by the function f by calculating a *differential quotient*, i. e., the ratio of the change in function value to an infinitesimal change in the argument. For the unit parabola $f(x) = x^2$ and infinitesimal dx, for example, the differential quotient is

$$\frac{f(x+dx) - f(x)}{dx} = \frac{(x+dx)^2 - x^2}{dx} = 2x + dx.$$

Leibniz explained that the infinitesimal summand dx can be omitted to determine the slope of the curve, in this case the unit parabola, as follows:

> Certainly, I agree with Euclid bk. 5, defin. 5, that only those homogeneous quantities are comparable of which one when multiplied by a number, that is, a finite number, can exceed the other. And I hold that any entities whose difference is not such a quantity are equal. (...) This is precisely what is meant by saying that the difference is smaller than any given.[3] (GM 5.322) (Leibniz as translated by Levey in [140], p. 147.)

---

1 "... and the rules of the finite are found to succeed in the infinite and *vice versa* the rules of the infinite are found to succeed in the finite ..." (own translation).

2 In the original Latin: "Nec refert an tales quantitates sint in rerum natura, sufficit enim fictione introduci ...".

3 In the original Latin: "Scilicet eas tantum homogeneas quantitates comparabiles esse, cum Euclide lib. 5. d[e]fin. 5 censeo, quarum una numero, sed finito, multiplicata, alteram superare potest. Et quae

This "generalized equality" allows for infinitesimal differences. Every ordinary finite quantity is, so to speak, wrapped in a fictitious cloud of infinitesimally deviating quantities, and this cloud can be stripped away again in the final result.

For Leibniz, unassignable quantities were useful fictions,

> ...since they allow economies of speech and thought in discovery as well as in demonstration. Nor is it necessary always to use inscribed or circumscribed figures, and to infer by reductio ad absurdum, and to show that the error is smaller than any assignable ([138], p. 128, as translated by Arthur in [9], p. 27).[4]

At the same time, Leibniz here hinted at a systematic connection between the abbreviated calculation with unassignable quantities and the more cumbersome, but since antiquity recognized *exhaustion method*, in which inscribed and circumscribed figures are used and reduced to absurdity. In a letter to Pinson, Leibniz formulated the justification of unassignable quantities as follows:

> Car au lieu de l'infini ou de l'infiniment petit, on prend des quantités aussi grandes et aussi petites qu'il faut pour que l'erreur soit moindre que l'erreur donnée, ... ([134], p. 494).[5]

In modern terms, this expresses the equivalence of nonstandard and standard analysis. The same results are obtained in both ways, once with a genuine infinitesimal calculus and once with the conventional limit calculus.

2.2 A modern translation of Leibniz's ideas

In the introduction to analysis outlined here, we transfer the ideas from Leibniz's infinitesimal calculus presented in the last section into the framework of a modern axiomatic theory.

- Instead of the historical concept of quantity, we use the axiomatic concept of *real number*. This also includes 0 and negative numbers.
- We understand analysis as an axiomatic theory that deals with *real numbers* and *sets* (collectively also referred to as *objects*). We identify functions with their graphs, i. e., we understand them as sets.
- In order to distinguish between assignable and unassignable quantities, we introduce the predicate *standard*, which is not defined, but belongs to the primitive no-

---

tali quantitate non differunt, aequalia esse statuo, quod etiam Archimedes sumsit, aliique post ipsum omnes. Et hoc ipsum est, quod dicitur differentiam esse data quavis minorem."

4 In the original Latin: "...cum loquendi cogitandique, ac proinde inveniendi pariter ac demonstrandi compendia praebeant, ne semper inscriptis vel circumscriptis uti, et ad absurdum ducere, et errorem assignabili quovis minorem ostendere necesse sit."

5 "For instead of the infinite or the infinitely small, we take quantities as great or as small as is needed for the error to be less than the given error, ..." (as translated in [8], footnote 53).

tions of the axiomatic theory (just like *real number* and *set*).[6] All objects that we can uniquely define in analysis (in particular, all numbers that can be explicitly specified) should be standard. On the other hand, nonstandard objects should not differ from standard objects in any way, which can be expressed without the predicate *standard*. This is guaranteed by the *axiom schema of transfer*, which takes over the role of Leibniz's principle quoted above, according to which rules of the finite are also applicable in the infinite and vice versa.

– Leibniz's statement on the existence of unassignable quantities ("it suffices that they be introduced by a fiction") becomes an axiom that postulates the existence of numbers that are greater than all standard numbers.

Remark. From a formalistic point of view, the ontological status of nonstandard numbers is in no way different from that of actual infinite sets. The latter also exist within the framework of a theory on the basis of axioms. While today the fiction of infinite sets is usually accepted without hesitation, this was not the case in Leibniz's time. Leibniz considered (bounded) infinite quantities to be useful fictions, but infinite collections (i. e., sets) to be contradictory because they violate the axiom "The whole is greater than its part" (see Section 5.2.2). In modern theories involving set theory, we can have both: infinite sets and infinite quantities (i. e., unlimited numbers).

2.3 Preliminary remarks on axiomatics

We develop analysis as an axiomatic theory. This means that we establish axioms and derive theorems from them. The concepts used in the axioms are the *primitive notions* of the theory. They are *not defined*, but implicitly determined by the axioms. For convenience, we can also introduce *defined concepts* and thus enrich the language.

We postulate the existence of a *domain of discourse* (also called *universe of discourse* or simply *universe*) in which the axioms (and thus also the derived theorems) hold. We consider this universe to be *actually infinite*. This means that we imagine that the universe is given in its entirety. Quantifying phrases such as "for all x ..." or "there exists x ..." always refer to the universe.

It is not possible to formulate or prove statements within an axiomatic theory that lead out of the theory, i. e., that use concepts that belong neither to the primitive notions nor to the defined concepts of the theory. The primitive notions of analysis are the unary predicates *real number* and *set*, the binary arithmetic operations + ("plus") and · ("times"), the constants 0 and 1, as well as the binary predicates < ("smaller than") and $\in$ ("element of"). In the theory of analysis presented here, the unary predicate *standard*

---

[6] The extension of the language is nevertheless "harmless" in the sense that the theory we develop will turn out to be conservative over ZF (see Section 2.11.1).

is added as a further primitive notion. Defined concepts can be, for example, further constants for numbers or sets or further predicates or operations.

In addition to the primitive notions and defined concepts, statements of analysis may in principle only contain variables (like x, y as placeholders for objects of the universe), the equality sign (=), logical connectives ("not", "and", "or", "if, then", "if and only if") and quantifications over variables (like "for all x" or "there exists x"). Variables that are not bound by a quantifier are called *free variables* (sometimes also referred to as *parameters*). *Open statements* contain free variables, *closed statements* do not. Closed statements are also called *sentences*.

Example. "$x + y = 0$" is an open statement with the free variables x and y. "For all real numbers x there exists a real number y with $x + y = 0$" is a closed statement. The variables x and y are bound here by the quantifiers "for all" and "there exists", respectively.

Variables are usually Latin lower or upper case letters, indexed if necessary or with other distinguishing characters. An exception to this rule are the Greek letters ε and δ, which are traditionally used in analysis as variables for real numbers. Upper case letters are mainly used as set variables. We do not introduce a formal language, but occasionally use the following logical symbols for abbreviation:
- the quantifiers $\forall$ ("for all"), $\exists$ ("there exists"),
- the connectives $\neg$ ("not"), $\wedge$ ("and"), $\vee$ ("or"), $\Rightarrow$ ("if ...then"), $\Leftrightarrow$ ("if and only if").

In the metalanguage, we use lower case Greek letters such as φ, χ, ψ as placeholders for statements. Unless otherwise stated, the statement $\varphi(x_1, \ldots, x_n)$ contains at most the free variables $x_1, \ldots, x_n$. Such a statement defines an *n-ary predicate*. Unary predicates are also called *properties*. If a statement may have other free variables (apart from those explicitly specified), we say that the statement "possibly has parameters".

If (due to the axioms still to be specified) for all $x_1, \ldots, x_n$ there is a unique y with $\varphi(x_1, \ldots, x_n, y)$, then the open statement $\varphi(x_1, \ldots, x_n, y)$ defines an *n-ary function on the universe* (also called *operation*) with the *input variables* $x_1, \ldots, x_n$ and the *output variable* y.[7] Special *operator symbols* are often introduced for operations, and one then writes, for example, $F(x_1, \ldots, x_n) = y$ (with the operator symbol F) instead of $\varphi(x_1, \ldots, x_n, y)$.

In the case $n = 0$, we have an operation without input variables, i. e., a statement $\varphi(y)$ that uniquely characterizes an object of the universe. A new symbol, a *constant*, can then be introduced to denote this object.

---

[7] Operations are not restricted with regard to the assignment of their input variables. This means that they are always defined on the *entire* universe. If we explicitly define some operations in the following only for a certain area of the universe (for example, for the area of sets or sets with certain additional properties), this means that we only use them in the context of this area. Outside their intended scope, their value can be arbitrarily defined (for example, as $\emptyset$).

Based on the previous explanations, our universe should contain real numbers and sets. However, we do not explicitly require that *all* objects of the universe are real numbers or sets. This gives us the option of introducing further primitive notions if necessary (for example, to save the effort of defining them in terms of set theory).

We also do not explicitly require that real numbers are *not sets*. This makes our theory compatible with the pure set theory ZFC, in which the real numbers are defined in terms of set theory. This means that our universe can also be understood as a ZFC universe.

2.4 Set theory in analysis

In contrast to many introductions to analysis, we will explicitly include the required set axioms in the axiomatics. Sets do not exist as naive, intuitively given collections, but because their existence follows from axioms. Only in this way can analysis be understood as an axiomatic theory. The existence of sets, like the existence of real numbers, is a *theoretical*, an axiomatically postulated existence.

2.4.1 The standard set axioms

In this section, we specify the standard set axioms that we will use in the following. First, we stipulate: A statement is called *internal* if it involves the predicate *standard* neither directly nor indirectly (i. e., via defined concepts). The other statements are called *external*. Thus, all statements of conventional analysis are internal.

Axiom (extensionality). *Two sets A and B are equal if they contain the same elements, i. e., if for all x,*

$$x \in A \quad \Leftrightarrow \quad x \in B.$$

Axiom (existence). *There exists a set that contains all real numbers.*[8]

Axiom schema (separation). *Let $\varphi(x, x_1, \ldots, x_n)$ be an* internal *statement. Then for all $x_1, \ldots, x_n$ and for any set E there exists a set E' that contains exactly those $x \in E$ for which $\varphi(x, x_1, \ldots, x_n)$ holds.*

The set E' (dependent on $x_1, \ldots, x_n$) is uniquely determined according to the axiom of extensionality and is denoted by $\{x \in E \mid \varphi(x, x_1, \ldots, x_n)\}$. The case $n = 0$ is permitted, which means that there are no other parameters in φ apart from x.

---

[8] Not to be confused with the axiom $\exists x\, x = x$, which is sometimes used in pure set theories as an "axiom of existence" (if the existence of a set does not follow from other axioms).

The constants $\mathbb{R}$ and $\emptyset$, the binary set operation $\cap$, the unary set operation $\bigcap$ and the binary predicate $\subseteq$ can be defined on the basis of these axioms.

According to the axiom of existence, there is a set E that contains all real numbers. By separation, there is then the set

$$\mathbb{R} := \{x \in E \mid x \text{ is a real number}\},$$

which is uniquely determined and independent of the concrete choice of E (by extensionality). From now on, we can write $x \in \mathbb{R}$ instead of "x is a real number".

There is also the uniquely determined set

$$\emptyset := \{x \in \mathbb{R} \mid x \neq x\}$$

(the *empty set*). Since the separating property $x \neq x$ is unfulfillable, $\emptyset$ contains no elements. Due to the axiom of extensionality, $\emptyset$ is the only set that contains no elements.

For any sets X, Y, there exist the sets

$$X \cap Y := \{x \in X \mid x \in Y\}$$

and

$$X \setminus Y := \{x \in X \mid x \notin Y\}$$

formed by separation. The set $X \cap Y$ is called the *intersection* of X and Y and contains exactly those elements that are contained in both sets X and Y, while $X \setminus Y$ is called the *set difference* of X and Y and contains exactly those elements of X that are *not* contained in Y.

A set whose elements are again sets is also called a *family of sets*. For any nonempty family of sets E and every $A \in E$, there exists the set

$$\bigcap E := \{x \in A \mid \forall X (X \in E \Rightarrow x \in X)\}$$

formed by separation. It is called the *intersection* of E. Due to the separating property, the definition does not depend on A. The set $\bigcap E$ is therefore well defined and contains exactly those elements that are contained in *all* sets of the family E.

For two sets X, Y, we write $X \subseteq Y$ (X is a *subset* of Y) if all elements of X are also elements of Y. More formally,

$$X \subseteq Y \quad :\Leftrightarrow \quad X, Y \text{ are sets and } \forall x \, (x \in X \Rightarrow x \in Y).$$

Furthermore, $X \subset Y$ means $X \subseteq Y$ and $X \neq Y$ ("X is *proper subset* of Y").

Remarks. 1. At this point, it can be discussed why there is no axiom that requires the existence of a set that contains all sets (analogous to the axiom of existence).

2. The axiom of separation is not a single axiom, but a so-called *axiom schema*. This means that for each internal statement there is an axiom according to this schema. Since statements are not objects of the postulated universe, but objects of our language, it is not possible to formulate an axiom of the type "For all statements $\varphi,\ldots$". Instead, the desired axioms are formulated as a schema. Although potentially an infinite number of axioms can be formed according to this schema, only a finite number of axioms, which must be specified in concrete terms, may be used in each proof. For the sake of simplicity, we sometimes say "axiom of separation" or "separation schema" instead of "axiom schema of separation" (analogous for other axiom schemas).
3. The fact that the separation schema is only formulated for internal statements will turn out to be essential later. First of all, it should be noted that this axiom schema is thus completely in accordance with the corresponding axiom schema of conventional analysis (where the predicate *standard* does not exist). This means that there are no restrictions compared to conventional analysis as far as the formation of sets is concerned.
4. A separation with external statements, for example,

$$\{x \in \mathbb{R} \mid x \text{ is standard}\},$$

is not covered by the axiom schema of separation and is called an *illegal set formation* (cf. [160], p. 1165). Such a "set term" is just as undefined as the "fraction" $\frac{1}{0}$. We will come back to this in Section 2.7.

Axiom (pairing). *For all x, y, there exists a set that contains exactly the elements x and y. It is uniquely determined according to the axiom of extensionality and is denoted by $\{x, y\}$ (the* pair *of x and y).*

Axiom (union). *For any family of sets E, there exists a set that contains exactly the elements of all sets of the family E. It is uniquely determined according to the axiom of extensionality and is denoted by $\bigcup E$ (the* union *of E).*

Axiom (power set). *For any set E, there is a set whose elements are exactly all possible subsets of E. It is uniquely determined according to the axiom of extensionality and is denoted by $\mathcal{P}(E)$ (the* power set *of E).*

The axiom of pairing is used to define a *singleton* (i. e., a set with a single element x) by

$$\{x\} := \{x, x\}.$$

Together with the axiom of union, sets with three, four, five, and so on elements can be formed:

$$\{x_1, x_2, x_3\} := \bigcup\{\{x_1, x_2\}, \{x_3\}\}.$$

In general,
$$\{x_1,\ldots,x_n\} := \bigcup\{\{x_1,\ldots,x_{n-1}\},\{x_n\}\}.$$

Here $\bigcup$ and $\mathcal{P}$ are unary set operations, while $\{.,\ldots,.\}$ (with n places between the braces) is an n-ary operation. The binary set operation $\cup$ can be defined with the axioms of pairing and union: For any sets X, Y, let
$$X \cup Y := \bigcup\{X,Y\}$$

(the *union of X and Y*). It contains exactly those elements that are contained in X or Y.

2.4.2 The set axioms not used

We will not use the following axioms of ZFC. However, they can be discussed ad hoc if required. We list them for the sake of completeness.

Axiom (regularity). *Every nonempty set contains an $\in$-minimal element, i. e., an element that has no element in common with the set.*

The axiom of regularity ensures the hierarchical structure of the universe of sets and prevents, for example, sets that contain themselves as elements (as well as other cyclic or infinite descending $\in$-chains). It is not needed for analysis. Its role in set theory is discussed in Section 5.4.2.

Axiom schema (replacement). *Let $\varphi(x,y,x_1,\ldots,x_n)$ be an internal statement. Then for all $x_1,\ldots,x_n$, if for every x there is a unique y with $\varphi(x,y,x_1,\ldots,x_n)$, then for any set X there is a set Y that contains exactly those y that satisfy $\varphi(x,y,x_1,\ldots,x_n)$ for some $x \in X$.*

With the axiom schema of replacement, sets of the form $\{F(x) \mid x \in X\}$ can be formed, where F denotes an operation (defined by an internal statement) and X is an arbitrary set. As mentioned at the end of Section 1.3.5, the axiom schema of replacement is used, for example, to prove the recursion theorem for operations. Here we are content with the recursion theorem for functions and can therefore manage with the weaker axiom schema of separation.

Axiom (infinity). *There is a set that contains $\emptyset$ as an element and for each of its elements x also the element $x \cup \{x\}$.*

In ZFC, the axiom of infinity ensures that there is a set (usually denoted by ω) that can be understood as the set of natural numbers (see Section 5.2.2). In an axiomatic introduction of real numbers, $\mathbb{N}$ is defined as the smallest inductive subset of $\mathbb{R}$. Therefore, the axiom of infinity is not needed to define the set $\mathbb{N}$.

Axiom (choice). *For each family of pairwise disjoint, nonempty sets, there exists a set that contains exactly one element of each set of the family.*

The axiom of choice is used in analysis, for example, to show that ε–δ-continuity follows from continuity in terms of limits of sequences. However, a weaker variant, the *axiom of countable choice* (see Section 5.4.10), is sufficient for this purpose.

The omission of the axiom of choice is significant with regard to the mathematical foundations, since the nonstandard extension set up in this way is *conservative* over ZF (see Section 2.11.1).

2.4.3 Cartesian products

Since we understand functions and relations as sets of ordered pairs, the question arises as to what ordered pairs should be within our theory. So far, we only know sets and real numbers as objects of our universe. We now have two options: Either we introduce *ordered pair* as a new primitive notion, i.e., as an *undefined* binary operation $(.,.)$ on our universe and demand the desired properties axiomatically, or we *define* such an operation using the concepts available so far and show that it has the desired properties. For the sake of simplicity, we first follow the first path and sketch the second path at the end of the section.

Axiom (OP). *For all x_1, x_2, y_1, y_2,*

$$(x_1, y_1) = (x_2, y_2) \Rightarrow x_1 = x_2 \wedge y_1 = y_2.$$

Axiom (CP). *For all sets X, Y, there exists a set Z that contains exactly the ordered pairs (x, y) with $x \in X$ and $y \in Y$.*

The set Z in CP is uniquely determined according to the axiom of extensionality. It is called the *Cartesian product* of X and Y and is denoted by $X \times Y$. Thus $\times$ is a binary set operation.

Similar to how we get from pairs to sets with three, four, five, and so on elements, we can start from ordered pairs (2-tuples) and define *n-tuples* for $n = 3, 4, 5$, and so on by

$$(x_1, \ldots, x_n) := ((x_1, \ldots, x_{n-1}), x_n).$$

Forming *n*-tuples is an *n*-ary operation.

As an alternative to the axiomatic introduction, we can *define* ordered pairs by

$$(x, y) := \{\{x, y\}, x\} \tag{2.1}$$

and show that they satisfy OP. Based on the definition, we have $(x, y) \in \mathcal{P}(\mathcal{P}(X \cup Y))$ for all $x \in X$ and $y \in Y$. Therefore, the Cartesian product of two sets X and Y can be defined by

$$X \times Y := \{z \in \mathcal{P}(\mathcal{P}(X \cup Y)) \mid \text{there exist } x \in X, y \in Y \text{ with } z = (x,y)\}. \qquad (2.2)$$

More detailed explanations can be found, for example, in [55], pp. 41–43.

2.4.4 Relations and functions

A *relation* from X to Y is a subset of $X \times Y$. For a relation $R \subseteq X \times Y$,

$$\text{dom}(R) := \{x \in X \mid \exists y\, (x,y) \in R\}$$

is the *domain* of R and

$$\text{ran}(R) := \{y \in Y \mid \exists x\, (x,y) \in R\}$$

is the *range* of R.

A relation $f \subseteq X \times Y$ is called a *mapping* or *function* from X to Y if for each $x \in X$ there is a unique $y \in Y$ with $(x,y) \in f$. In this case, we write $f : X \to Y$ and usually $f(x) = y$ instead of $(x,y) \in f$.

Remark. Set formations of the type $\{f(x) \mid x \in A\}$ with a function f and $A \subseteq \text{dom}(f)$ are possible by separation as

$$\{y \in \text{ran}(f) \mid \exists x \in A\, f(x) = y\}.$$

The replacement schema is not required for this.

A mapping $f : X \to Y$ is called
- *injective* if for all $x_1, x_2 \in X$, $f(x_1) = f(x_2)$ implies $x_1 = x_2$,
- *surjective* if $\text{ran}(f) = Y$,
- *bijective* if it is injective and surjective.

Definition 1. Two sets X and Y are called *equipotent* ($X \sim Y$) if there is a bijective mapping $f : X \to Y$.

It is easy to see that the binary set predicate $\sim$ is *reflexive* ($X \sim X$ holds for any set X), *symmetric* (if $X \sim Y$, then $Y \sim X$), and *transitive* (if $X \sim Y$ and $Y \sim Z$, then $X \sim Z$).

2.5 The Archimedean field of real numbers

The field and order axioms are formulated exactly as in conventional analysis and the usual theorems are derived from them. We state the axioms here without further comments and only go into more detail about the embedding of the natural numbers. Once we have defined the set $\mathbb{N}$, we can formulate the Archimedean axiom as usual.

The completeness axiom is formulated in conventional analysis, for example, as the least-upper-bound property or (together with the Archimedean axiom) via the convergence of all Cauchy sequences. In the axiomatics presented here, the completeness of $\mathbb{R}$ will result from the *standard part axiom*. We introduce this axiom in Section 2.6.3 and obtain the least-upper-bound property as a theorem.

As usual, we define $x \le y$ by $x < y \lor x = y$, $x > y$ by $y < x$, and $x \ge y$ by $y \le x$.

2.5.1 The field axioms

Axiom.
1. $(x + y) + z = x + (y + z)$ for all $x, y, z \in \mathbb{R}$.
2. $x + y = y + x$ for all $x, y \in \mathbb{R}$.
3. $0 \in \mathbb{R}$.
4. $x + 0 = x$ for all $x \in \mathbb{R}$.
5. For each $x \in \mathbb{R}$, there exists $y \in \mathbb{R}$ such that $x + y = 0$.
6. $(x \cdot y) \cdot z = x \cdot (y \cdot z)$ for all $x, y, z \in \mathbb{R}$.
7. $x \cdot y = y \cdot x$ for all $x, y \in \mathbb{R}$.
8. $1 \in \mathbb{R}$.
9. $1 \ne 0$.
10. $x \cdot 1 = x$ for all $x \in \mathbb{R}$.
11. For each $x \in \mathbb{R} \setminus \{0\}$, there exists $y \in \mathbb{R}$ such that $x \cdot y = 1$.
12. $x \cdot (y + z) = (x \cdot y) + (x \cdot z)$ for all $x, y, z \in \mathbb{R}$.

2.5.2 The order axioms

Axiom.
1. For all $x \in \mathbb{R}$, exactly one of the three relations holds:

$$x < 0, \quad x = 0, \quad x > 0.$$

2. For all $x, y \in \mathbb{R}$, if $x > 0$ and $y > 0$, then $x + y > 0$.
3. For all $x, y \in \mathbb{R}$, if $x > 0$ and $y > 0$, then $x \cdot y > 0$.

2.5.3 Embedding the natural numbers

When embedding the natural numbers, it makes sense to address the problem discussed in Section 1.3.1: Why cannot we simply define $\mathbb{N}$ as the set of all finite sums of ones? We would have to define within the theory what a finite sum of ones is, i. e., a term of the form

$$\underbrace{1 + \cdots + 1}_{n \text{ summands}},$$

where *n* is a natural number. We would therefore already need the concept of a natural number in relation to the summands of a term. However, terms are not objects of the postulated universe of analysis, but objects of the *language of analysis*, more precisely, character strings with a certain structure. They can *denote* objects of the universe, but do not themselves *belong* to the universe. We therefore cannot speak about terms or prove statements about terms within the theory.

We must therefore distinguish the colloquial natural numbers, which we use to speak about terms, formulas, and proofs, as well as everyday objects, from the natural numbers of an axiomatic theory such as analysis. In this context, the numbers of the first type are called natural numbers of the *metalanguage* (short: *metanumbers*) and the numbers of the second type natural numbers of the *object language* (short: *object numbers*).

The following relationship exists between metanumbers and object numbers: For every metanumber n, there is an object number that is represented by a sum term with n ones. This does not mean that, conversely, every object number can be represented by a sum term of ones.

The set $\mathbb{N}$ of object numbers is defined as in Section 1.3.2 as the smallest inductive subset of $\mathbb{R}$, i. e.,

$$\mathbb{N} := \bigcap \{E \in \mathcal{P}(\mathbb{R}) \mid E \text{ is inductive}\},$$

where a set $E \subseteq \mathbb{R}$ is called *inductive*, provided that

$$1 \in E \quad \text{and} \quad \text{for all } n, \text{ if } n \in E \text{ then } n + 1 \in E.$$

2.5.4 Definitions and proofs in the metalanguage

Just as a distinction must be made between the natural numbers of the metalanguage and those of the object language, a distinction must also be made between the metalanguage and object language versions for inductive proofs and recursive definitions.

To take this distinction into account visually, we use the symbols 1, 2, 3, ... and the variable $\mathfrak{n}$ for the metanumbers, and the symbols 1, 2, 3, ... and the variable n for the object numbers.

We consider the metanumbers to be *potentially infinite*. This means that we assume that we can continue the counting sequence 1, 2, 3, ... as far as we want (as far as we need it in each case), but we do not claim that the metanumbers are given in their entirety. In contrast, we have proved (i. e., deduced from the axioms) that in the postulated universe of analysis the *actual infinite* set $\mathbb{N}$ of all object numbers exists.

Recursive definitions in the metalanguage

One can define concepts $F(n)$ of the object language *recursively* by specifying (in the metalanguage) what the definition of $F(1)$ is and how to obtain the definition of $F(n + 1)$ from the definition of $F(n)$.

Note that we do not understand F here as an infinite set of all ordered pairs $(n, F(n))$ (i.e., as a completed infinity), but as an operative instruction on how we can construct $F(1), F(2), F(3), \ldots$ in principle to any extent (as far as we need it in each case).

To clarify once again: The concrete definitions of the concepts $F(1), F(2), F(3), \ldots$ belong to the *object language*, but the recursive instruction on how to construct these definitions one after the other belongs to the *metalanguage*. That is why we call this instruction a *recursive definition in the metalanguage*.

As examples of recursive definitions in the metalanguage, we have already seen the definition of the n-ary operations $\{x_1, \ldots, x_n\}$ and $(x_1, \ldots, x_n)$ in Section 2.4.1.

Induction in the metalanguage

If a metalanguage statement $A(n)$ is valid for $n = 1$ and if for any given n one can infer from $A(n)$ to $A(n + 1)$, then $A(n)$ is valid for any n.

Again, note that this is not to be understood as a statement about an infinite set of all metanumbers, but in such a way that as you progress in the potentially infinite counting sequence $1, 2, 3, \ldots$ you will only come across numbers n for which $A(n)$ holds.

For example, metalanguage induction can be used to show that two n-tuples are equal if and only if they agree in each component. This is not a statement of the object language, but a statement *about* statements of the object language and thus a statement of the metalanguage. Strictly speaking, the statement reads: "For any metanumber n, the following statement of the object language is valid (i.e., a theorem): Two n-tuples are equal if and only if they agree in each component." Metalanguage induction must therefore be used to prove this.

The justification for the induction principle in the metalanguage is that for each n one has the conclusion chain

$$A(1) \Rightarrow A(2) \Rightarrow A(3) \Rightarrow \cdots \Rightarrow A(n)$$

and thus a proof for $A(n)$.

Such a justification is not available for induction over the object numbers (which form an actual infinite set). Instead, the corresponding principles of definition and proof must first be proven on the basis of the axioms.

2.5.5 Again: Mathematical induction and recursive definitions

Based on the definition of $\mathbb{N}$, we have the following

Theorem 1 (mathematical induction). *Let $\varphi(n)$ be an internal statement (possibly with parameters). If $\varphi(1)$ holds and if for any $n \in \mathbb{N}$, $\varphi(n)$ implies $\varphi(n+1)$, then $\varphi(n)$ holds for all $n \in \mathbb{N}$.*

Proof. According to the axiom schema of separation, there exists the set $E := \{n \in \mathbb{N} \mid \varphi(n)\}$. By assumption, E is inductive and therefore $\mathbb{N} \subseteq E$. Hence, $\varphi(n)$ holds for all $n \in \mathbb{N}$. □

To show that an internal statement $\varphi(n)$ is valid for all $n \in \mathbb{N}$, the proof by mathematical induction proceeds according to the following schema (which is analogous to induction in the metalanguage):

Base case: Show $\varphi(1)$.
Induction hypothesis: Suppose $\varphi(n)$ for an arbitrary $n \in \mathbb{N}$.
Induction step: Show that $\varphi(n)$ implies $\varphi(n+1)$.

Sometimes it is convenient to start the induction not at 1, but at 0 or more generally at some number $n_0 \in \mathbb{N}_0$, and thus show that an internal statement $\varphi(n)$ holds for all $n \in \mathbb{N}_0, n \geq n_0$.

Remarks. 1. Analogous to the axiom schema of separation, Theorem 1 is, strictly speaking, not a single theorem, but a theorem *schema*. This means that for each internal statement $\varphi(n)$ there is a corresponding theorem, potentially an infinite number of theorems.
2. The condition that $\varphi(n)$ must be an internal statement does not imply any restriction compared to conventional analysis, where the predicate *standard* does not exist and therefore no external statements.

By mathematical induction, it can be shown that the elements of $\mathbb{N}$ have the properties familiar from dealing with naive natural numbers, e. g., that
- sums and products of natural numbers are natural numbers again,
- all natural numbers are positive,
- all natural numbers except 1 have a natural number as their predecessor.

The proofs are straightforward. To show the first assertion for $m, n \in \mathbb{N}$, assume $m \in \mathbb{N}$ is fixed (but arbitrary) and conclude by induction on n. For the second assertion, note that the base case $1 > 0$ follows from the axioms of an ordered field (see, for example, [188], p. 8).

With some experience in induction proofs, it seems tedious and unnecessary to go through each proof in detail (and hardly any analysis course would do that), but it is important to realize that these proofs are necessary in principle and that induction proofs remain valid regardless of whether there are nonstandard numbers in $\mathbb{N}$ or not. We show some important examples.

2.5 The Archimedean field of real numbers

Theorem 2 (well-ordering principle). *Every nonempty subset of $\mathbb{N}$ contains a least element.*

Proof. Let A be a subset of $\mathbb{N}$ that contains no least element. We have to show that A is empty. Let $\varphi(n)$ be the following internal statement (with parameter A):

$$k \notin A \quad \text{for all } k \in \mathbb{N} \text{ with } k \leq n.$$

Then $\varphi(1)$ holds because otherwise 1 would be the least element of A. If $\varphi(n)$ holds for any $n \in \mathbb{N}$, then $\varphi(n+1)$ also holds because otherwise $n+1$ would be the least element of A. Therefore, according to Theorem 1, $\varphi(n)$ holds for all $n \in \mathbb{N}$. Hence A must be empty. □

Theorem 3 (recursion theorem for functions). *Let A be a nonempty set, $a \in A$, and $F: A \to A$ a function. Then there exists a unique function $f: \mathbb{N} \to A$ such that*
1. $f(1) = a$,
2. $f(n+1) = F(f(n))$ for all $n \in \mathbb{N}$.

Proof. (Uniqueness) Let $g: \mathbb{N} \to A$ be another function with the properties given in the theorem. Then $f(1) = a = g(1)$ and

$$f(n) = g(n) \Rightarrow f(n+1) = F(f(n)) = F(g(n)) = g(n+1)$$

for all $n \in \mathbb{N}$. By Theorem 1, $f = g$ follows.

(Existence) Let $\varphi(h)$ be the following internal statement (with parameters A, F and a):

$$h \subseteq \mathbb{N} \times A$$
$$\text{and} \quad (1, a) \in h$$
$$\text{and} \quad \text{for all } (n, b) \in \mathbb{N} \times A, \text{ if } (n, b) \in h \text{ then } (n+1, F(b)) \in h.$$

This statement defines a *property* (dependent on A, F, a). Let us call a set with this property a φ-set. Then $\mathbb{N} \times A$ is a φ-set, and any intersection of φ-sets is also a φ-set again. Therefore, the intersection

$$f := \bigcap \{h \in \mathcal{P}(\mathbb{N} \times A) \mid \varphi(h)\}$$

is the *smallest* φ-set. It contains exactly those elements that are contained in *every* φ-set. As a subset of $\mathbb{N} \times A$, f is a binary relation. By construction, $\text{dom}(f) \subseteq \mathbb{N}$ is inductive, hence $\text{dom}(f) = \mathbb{N}$. It remains to show that f is *right-unique*, i.e., that for all $n \in \mathbb{N}$ there is a unique $b \in A$ with $(n, b) \in f$. This is done by induction.

Base case ($n = 1$): Since f is a φ-set, we have $(1, a) \in f$. If there were $a' \in A$ with $a' \neq a$ and $(1, a') \in f$, then $f \setminus \{(1, a')\}$ would still be a φ-set, and f would not be the smallest φ-set.

Induction step: As induction hypothesis, we assume for an arbitrary $n \in \mathbb{N}$ that there is a unique $b \in A$ with $(n, b) \in f$. Since f is a φ-set, we then have $(n+1, F(b)) \in f$. If there were $b' \in A$ with $b' \neq F(b)$ and $(n+1, b') \in f$, then $f \setminus \{(n+1, b')\}$ would still be a φ-set and f would not be the smallest φ-set. □

The recursion theorem enables recursive definitions, for example, the definition of the sequence of partial sums for a given sequence of real numbers, as described in Section 1.3.5.

2.5.6 Again: Finite sets

In the following we write $\{1, \ldots, n\}$ (with $n \in \mathbb{N}_0$) as an abbreviation for $\{k \in \mathbb{N} \mid 1 \leq k \leq n\}$. This is the empty set if and only if $n = 0$.

Definition 2. A set A is called *finite* if there is $n \in \mathbb{N}_0$ such that

$$A \sim \{1, \ldots, n\}.$$

Sets that are not finite are called *infinite*.

This is a common definition that we already quoted in Section 1.3.6. Statements about finite sets are usually proved by induction on the number of their elements (which we can define after the next theorem). We then show two more examples.

Theorem 4. For all $m, n \in \mathbb{N}_0$,

$$\{1, \ldots, m\} \sim \{1, \ldots, n\} \Rightarrow m = n.$$

Proof. We show by induction on m that the following statement holds for all $m \in \mathbb{N}_0$:

$$\forall n \in \mathbb{N}_0 \, (\{1, \ldots, m\} \sim \{1, \ldots, n\} \Rightarrow m = n). \tag{2.3}$$

Base case ($m = 0$): From $\{1, \ldots, n\} \sim \{1, \ldots, m\} = \emptyset$ follows $\{1, \ldots, n\} = \emptyset$ and hence $n = 0$.

Induction hypothesis: (2.3) holds for an arbitrary $m \in \mathbb{N}_0$.

Induction step: Let $f : \{1, \ldots, m+1\} \to \{1, \ldots, n\}$ be bijective. Then n is not equal to 0 and therefore the successor of a number $n' \in \mathbb{N}_0$. Let $g : \{1, \ldots, n\} \to \{1, \ldots, n\}$ be defined by $g(n) = f(m+1)$, $g(f(m+1)) = n$ and $g(i) = i$ otherwise. The mapping g only swaps n and $f(m+1)$ (if they are different) and is therefore bijective. Then $g \circ f : \{1, \ldots, m+1\} \to \{1, \ldots, n\}$ is bijective and maps (by definition of g) $m+1$ to n. Therefore, $g \circ f \setminus \{(m+1, n)\}$ is a bijective mapping from $\{1, \ldots, m\}$ to $\{1, \ldots, n'\}$. By induction hypothesis, $m = n'$ and hence $m + 1 = n' + 1 = n$. □

Theorem 4 enables the following

Definition 3. Let A be a finite set and $n \in \mathbb{N}_0$ the uniquely determined number with $A \sim \{1, \ldots, n\}$. Then n is called the *cardinality* or *number of elements* of A. It is denoted by $|A|$.

Theorem 5. *If A is a finite set and $B \subset A$, then B is also finite and $|B| < |A|$.*

Proof. It is sufficient to show the assertion for sets $A = \{1, \ldots, n\}$, $n \in \mathbb{N}_0$.
  Base case ($n = 0$): In this case we have $A = \emptyset$, and there is nothing to show.
  Induction hypothesis: The assertion holds for an arbitrary $n \in \mathbb{N}_0$.
  Induction step: Let $A := \{1, \ldots, n+1\}$ and $B \subset A$. Let $n+1 \notin B$ be assumed first. Then $B \subseteq \{1, \ldots, n\}$. In the case of equality it follows: B is finite and $|B| = n < n+1 = |A|$. In the case of "⊂", it follows by induction hypothesis: B is finite and $|B| < n < n+1 = |A|$. Now let $n+1 \in B$ be assumed. Let m be the least element of $A \setminus B$ and $B' := B \setminus \{n+1\} \cup \{m\}$. Then, according to the consideration just made, the assertion holds for B' and thus also for B, because if B' is finite, then B is also finite and $|B| = |B'| < |A|$. □

Theorem 6. *Every finite nonempty subset of $\mathbb{R}$ contains a smallest element (minimum) and a largest element (maximum).*

Proof. For a singleton $\{a_1\} \subseteq \mathbb{R}$, a_1 is both the minimum and the maximum. Now let $A := \{a_1, \ldots, a_{n+1}\} \subseteq \mathbb{R}$ be a set with $n+1$ elements, and (by induction hypothesis) let a_{i_1} be the minimum and a_{i_2} the maximum of the set $\{a_1, \ldots, a_n\}$. Then $\min(a_{i_1}, a_{n+1})$ is the minimum and $\max(a_{i_2}, a_{n+1})$ is the maximum of A. □

2.5.7 The Archimedean axiom

The Archimedean axiom is formulated as in conventional analysis, and the usual conclusions are derived from it.

Axiom (Archimedean axiom). *For each real number x, there is a natural number n with $n > x$.*

An equivalent, frequently used reformulation is: For any two positive real numbers x and y, there is a natural number n with $n \cdot x > y$.

2.5.8 Interim conclusion

Everything we have explained so far is used in the same way in conventional analysis, even if it is usually not explicitly mentioned there. Apart from the additional predicate *standard* (which we have not yet used), there is therefore no difference in axiomatics between conventional and nonstandard analysis up to this point.

2.6 The nonstandard axioms of analysis

2.6.1 Axiom schema of transfer

We now come for the first time to an axiom that does not exist in conventional analysis. Accordingly, the predicate *standard* will appear explicitly for the first time. We use the relativized quantifiers $\forall^{st} x$ and $\exists^{st} x$ for "For all standard x" and "There exists a standard x", respectively.[9]

The axiom schema of transfer is intended to ensure that the nonstandard objects do not differ from the standard objects in any property that can be formulated without the predicate *standard*. We therefore demand for any internal statement $\varphi(x)$,

$$\forall^{st} x\, \varphi(x) \Rightarrow \forall x\, \varphi(x). \tag{2.4}$$

Trivially, the reverse direction "$\Leftarrow$" is also valid.
If we apply (2.4) to $\neg\varphi$, we get the dual variant

$$\exists x\, \varphi(x) \Rightarrow \exists^{st} x\, \varphi(x). \tag{2.5}$$

The reverse direction is trivial again.

It follows from (2.5) that every conventionally defined object is standard because the definition is expressed by an internal statement $\varphi(x)$ that holds for a unique x. According to (2.5), this x must be standard. This means, for example, that the real numbers 10^{10}, $-\frac{13}{5}$, $\sqrt{2}$, π, the sets $\mathbb{N}$, $\mathbb{Z}$, $\mathbb{Q}$, $\mathbb{R}$, and the functions sin, cos, exp are standard.

In the final version of the transfer axiom, the statement $\varphi(x)$ is allowed to have parameters $x_1, \ldots, x_n$, but their range is restricted to standard objects. They are therefore called *standard parameters*.

Axiom schema (transfer). *Let $\varphi(x, x_1, \ldots, x_n)$ be an internal statement without further free variables ($n = 0$ allowed). Then for all standard $x_1, \ldots, x_n$,*

$$\forall^{st} x\, \varphi(x, x_1, \ldots, x_n) \Rightarrow \forall x\, \varphi(x, x_1, \ldots, x_n).$$

Analogous to (2.5), we have in this case for all standard $x_1, \ldots, x_n$,

$$\exists x\, \varphi(x, x_1, \ldots, x_n) \Rightarrow \exists^{st} x\, \varphi(x, x_1, \ldots, x_n).$$

Consequently, everything that can be defined by an internal statement with standard parameters is standard. For example,
- $x + y$ and $x \cdot y$ are standard real numbers, provided that x and y are standard real numbers,

---

[9] In principle, we can eliminate these new symbols by writing $\forall x\, (x \text{ standard} \Rightarrow \ldots)$ instead of $\forall^{st} x \ldots$ and $\exists x\, (x \text{ standard} \wedge \ldots)$ instead of $\exists^{st} x \ldots$.

- $A \cup B, A \cap B, A \setminus B, A \times B, \bigcup A, \bigcap A, \mathcal{P}(A)$ are standard sets, provided that A and B are standard sets,
- the set $\{x, y\}$ is standard, provided that x and y are standard (analogous for sets with three, four, five, etc., elements),
- the ordered pair (x, y) is standard, if and only if x and y are standard (analogous for triples, quadruples, quintuples, etc.),
- a standard function f has standard values $f(x)$ for all standard $x \in \text{dom}(f)$,
- the domain and the range of a standard function are standard sets.

Remarks. 1. The transfer axiom can be applied several times in succession if required. For example, if an internal statement $\varphi(x_1, \ldots, x_n)$ holds for all standard $x_1, \ldots, x_n$, then applying the transfer axiom n times yields $\varphi(x_1, \ldots, x_n)$ for all $x_1, \ldots, x_n$.

2. The transfer axiom is applicable to internal statements (with standard parameters). Without these prerequisites, application of the axiom is not permitted and is called an *illegal transfer* (cf. [160], p. 1166). If the transfer axiom also applied to external statements, the additional predicate *standard* would be senseless because it would immediately follow from the statement "x is standard" that all objects in the universe are standard.

According to the transfer axiom, all sets that can be defined by internal statements (for example, $\mathbb{N}$ or $\mathbb{R}$) are standard. However, this does *not* mean that all elements of these sets are also standard. It is precisely the purpose of the extended theory to have nonstandard numbers available in $\mathbb{R}$.

It is therefore at least possible (we cannot say more at the moment) that standard sets also contain nonstandard elements. Due to the transfer axiom, however, it is clear that a standard set is already uniquely determined by its standard elements (even if there are also nonstandard elements in it). To compare two standard sets, it is therefore sufficient to look at the standard elements, as the following theorem confirms.

Theorem 7. 1. To show $A \subseteq B$ for two standard sets A, B, it is sufficient to show that B contains all standard elements of A. As a formula,

$$\forall^{st} x \, (x \in A \Rightarrow x \in B) \quad \Rightarrow \quad A \subseteq B.$$

2. To show $A = B$ for two standard sets A, B, it is sufficient to show that A and B contain the same standard elements. As a formula,

$$\forall^{st} x \, (x \in A \Leftrightarrow x \in B) \quad \Rightarrow \quad A = B.$$

Proof. The statement $(x \in A \Rightarrow x \in B)$ is internal with standard parameters A and B. If it holds for all standard x, then (by transfer) it holds for all x, which means $A \subseteq B$. The second assertion can be shown analogously. □

2.6.2 Idealization axiom for real numbers

So far, we do not know whether there are any nonstandard objects in our postulated universe. The transfer axiom merely states that every object that we can conventionally define is standard and that nonstandard objects (if they exist) are completely unremarkable with respect to conventional analysis. If all objects in the universe were standard, then the extension of the theory by the predicate *standard* would be trivial (and uninteresting). If the question of the existence of nonstandard objects remained open, the new predicate would be useless.

At this point, we have to make a decision. Do we want nonstandard objects to exist? Then we must demand this by means of a suitable axiom. If we dispense with such an axiom, the predicate *standard* remains ineffective. Since nonstandard objects open up the possibility of infinitesimal numbers, we opt for an axiom that ensures the existence of nonstandard objects. Compared to the idealization schema in IST (see Section 3.5.3), the weaker idealization axiom for real numbers is sufficient here.

Axiom (idealization for real numbers). *There exists $x \in \mathbb{R}$ such that $x > y$ for all standard $y \in \mathbb{R}$.*

Since all numbers definable in conventional analysis are standard, numbers larger than all standard numbers would be "infinitely large" in an intuitive sense. As we noted in the introduction (Section 1.1), we prefer to call them *unlimited* in order to avoid confusion with the set-theoretical term *infinite*. The following terms have become established in nonstandard analysis.

Definition 4. 1. A number $x \in \mathbb{R}$ is said to be
   - *limited* if there exists a standard $y \in \mathbb{R}$ with $|x| \leq y$,
   - *unlimited* ($|x| \gg 1$) if x is not limited, i. e., if $|x| > y$ for all standard $y \in \mathbb{R}$,
   - *infinitely small* or *infinitesimal* ($x \approx 0$) if $|x| < y$ for all standard $y \in \mathbb{R}, y > 0$.
2. Numbers $x, y \in \mathbb{R}$ are said to be *infinitely close* ($x \approx y$) if $x - y \approx 0$.

Intuitively plausible calculation rules for limited, unlimited, and infinitesimal numbers can be easily verified using their definitions. The most important rules are summarized in the following theorem.

Theorem 8. *For all $x, y \in \mathbb{R}$,*
1. *if x is not infinitesimal, then x^{-1} is limited,*
2. *if x is unlimited, then x^{-1} is infinitesimal,*
3. *if $x \neq 0$ is infinitesimal, then x^{-1} is unlimited,*
4. *if x, y are limited, then $x + y$ and $x \cdot y$ are limited,*
5. *if x, y are infinitesimal, then $x + y$ and $x \cdot y$ are infinitesimal,*
6. *if x is positive (negative) and unlimited and y is limited, then $x + y$ is positive (negative) and unlimited,*

7. if x is unlimited and y is not infinitesimal, then $x \cdot y$ is unlimited,
8. if x is limited and y infinitesimal, then $x \cdot y$ is infinitesimal.

Proof. Exercise. □

From Definition 4 and the calculation rules for infinitesimal numbers, it follows that $\approx$ is reflexive, symmetric, and transitive, and thus has the typical properties of an equivalence relation on $\mathbb{R}$.[10]

Note that the implication $a \approx b \Rightarrow ac \approx bc$ is in general only valid if c is limited.

According to the idealization axiom for real numbers, there are unlimited numbers and (as their reciprocals) also nonzero infinitesimals in $\mathbb{R}$. Due to the Archimedean axiom, there are also unlimited natural numbers, and we obtain the following theorem.

Theorem 9. *For $x \in \mathbb{R}$,*
- *x is limited if and only if there is a standard $n \in \mathbb{N}$ with $|x| \leq n$,*
- *x is unlimited if and only if $|x| > n$ for all standard $n \in \mathbb{N}$,*
- *x is infinitesimal if and only if $|x| < \frac{1}{n}$ for all standard $n \in \mathbb{N}$.*

Zero is a special infinitesimal number due to

Theorem 10. *Zero is the only infinitesimal standard number.*

Proof. Let $x \in \mathbb{R}$ be infinitesimal and standard. By Definition 4, we have $\forall^{st} y > 0$ ($|x| < y$). This is an internal statement with standard parameter x. Therefore, the transfer axiom is applicable and $\forall y > 0$ ($|x| < y$) follows. Hence $x = 0$. □

Theorem 10 implies that two standard numbers that are infinitely close to each other must be equal.

Remark. The idea that there can be unlimited numbers in $\mathbb{N}$ and thus natural numbers greater than any sum of ones is unusual, and we will return to what this means for the concept of natural numbers in Section 2.7 (and in more detail in Section 5.5). On the other hand, this idea is very fruitful for understanding theoretical mathematics and its foundations. While the distinction between metanumbers and object numbers in standard analysis may seem like a pedantic formality without deeper meaning, the possibility of nonstandard extensions of the theory is striking proof that this distinction is absolutely necessary and—as further investigation will show—even very useful.

---

10 We cannot say that $\approx$ *is* an equivalence relation on $\mathbb{R}$ because $\{(x, y) \in \mathbb{R} \times \mathbb{R} \mid x \approx y\}$ would be an illegal set formation.

2.6.3 The standard part axiom and the completeness of real numbers

The standard part axiom can be understood as the nonstandard version of the completeness axiom. The idea of the conventional completeness axiom is that problems that can only be solved with arbitrary precision in the rational numbers should have an exact solution in the real numbers. A simple example is the task of solving the equation $x^2 = 2$. More generally, this requirement for the real numbers is expressed in the convergence of all Cauchy sequences or in the least-upper-bound property.

The axioms of the real numbers listed so far (field axioms, order axioms, Archimedean axiom) all hold in the rational numbers as well. Nevertheless, due to the availability of nonstandard numbers, we are in some respects one step further than in standard analysis because we can, for example, solve the equation $x^2 = 2$ with rational nonstandard numbers even with infinite precision. For example, choose an arbitrary unlimited $n \in \mathbb{N}$ and the smallest $m \in \mathbb{N}$ for which $(\frac{m}{n})^2 \geq 2$. Then $(\frac{m-1}{n})^2$ is still smaller than 2 and $(\frac{m}{n})^2 \approx 2$.[11]

If we now knew that there is a *standard* real number s infinitely close to $\frac{m}{n}$, then we would have an *exact* solution with s because $s \approx \frac{m}{n}$ implies $s^2 \approx (\frac{m}{n})^2 \approx 2$ (since $\frac{m}{n}$ is limited) and thus (since s^2 is standard) $s^2 = 2$. In fact, the completeness of $\mathbb{R}$ is expressed by the following axiom.

Axiom (standard part). *For all limited $x \in \mathbb{R}$, there is a standard $y \in \mathbb{R}$ with $x \approx y$.*

The number y in the standard part axiom is uniquely determined (because $\approx$ is transitive and two infinitely close standard numbers are equal). It is called the *standard part* of x and is denoted by $\mathrm{stp}(x)$. The following rules for calculating with standard parts can be easily verified using the definition.

Theorem 11. *For all limited $x, y \in \mathbb{R}$,*
1. *$x + y$ is limited and $\mathrm{stp}(x + y) = \mathrm{stp}(x) + \mathrm{stp}(y)$,*
2. *$x \cdot y$ is limited and $\mathrm{stp}(x \cdot y) = \mathrm{stp}(x) \cdot \mathrm{stp}(y)$,*
3. *$\frac{x}{y}$ is limited and $\mathrm{stp}(\frac{x}{y}) = \frac{\mathrm{stp}(x)}{\mathrm{stp}(y)}$, provided that y is not infinitesimal,*
4. *$x \leq y \Rightarrow \mathrm{stp}(x) \leq \mathrm{stp}(y)$.*

Proof. We show statements 1 and 4 as examples.

1. If x, y are limited, then $x + y$ is also limited (according to Theorem 8, point 4). Furthermore, there are infinitesimals $\tilde{x}, \tilde{y}$ with $x = \mathrm{stp}(x) + \tilde{x}$ and $y = \mathrm{stp}(y) + \tilde{y}$. Then

$$x + y = \mathrm{stp}(x) + \mathrm{stp}(y) + \tilde{x} + \tilde{y}.$$

Since $\mathrm{stp}(x) + \mathrm{stp}(y)$ is standard and $\tilde{x} + \tilde{y}$ is infinitesimal, the assertion follows.

---

[11] $\frac{(m-1)^2}{n^2} < 2 \leq \frac{m^2}{n^2}$ implies $0 \leq \frac{m^2}{n^2} - 2 < \frac{m^2}{n^2} - \frac{(m-1)^2}{n^2} = \frac{2m-1}{n^2} = \frac{1}{n}(2\frac{m}{n} - \frac{1}{n}) \approx 0$, since $\frac{m}{n}$ is limited and $n \gg 1$.

4. Let $x \leq y$. Now suppose $\text{stp}(x) > \text{stp}(y)$. Then $x - y \leq 0$, but $\text{stp}(x - y) = \text{stp}(x) - \text{stp}(y) > 0$. This means that the distance between $x - y$ and $\text{stp}(x - y)$ would be greater than or equal to a positive standard number, i.e., not infinitesimal, in contradiction to the definition of the standard part. Therefore, the assumption is incorrect and $\text{stp}(x) \leq \text{stp}(y)$ follows. □

The conventional description of completeness (i.e., the least-upper-bound property) follows from the standard part axiom (and the Archimedean axiom). The terms *upper bound* and *supremum* (*least upper bound*) are defined as usual, i.e., by internal statements.

Theorem 12 (least-upper-bound property). *Every nonempty subset of $\mathbb{R}$ that is bounded above has a least upper bound (supremum) in $\mathbb{R}$. The supremum of a set A is denoted by $\sup A$.*

Proof. We show that the assertion holds for standard sets. By transfer, it then holds in general.

Let $A \subseteq \mathbb{R}$ be standard, nonempty, and bounded above. According to the transfer axiom, there is a *standard* $a \in A$ and a *standard* upper bound r of A. Let $\alpha \approx 0$, $\alpha > 0$, and

$$M := \{j \in \mathbb{N} \mid a + j\alpha \text{ is upper bound of } A\}.$$

By the Archimedean axiom, there is $n \in \mathbb{N}$ with $a + n\alpha > r$. Thus M is a nonempty subset of $\mathbb{N}$ and (since $\mathbb{N}$ is well ordered) contains a minimum m. The number $a + m\alpha$ is limited (for example smaller than the standard number $r + 1$). Define $s := \text{stp}(a + m\alpha)$ (see Figure 2.1).

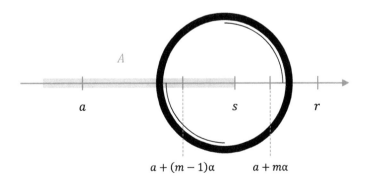

Figure 2.1: Nonstandard proof of the least-upper-bound property with unlimited magnification at s.

We show that s is an upper bound of A. By definition of m for all $x \in A$, $x \leq a + m\alpha$, hence (by Theorem 11, point 4) $\text{stp}(x) \leq s$. Hence for all standard x,

$$x \in A \Rightarrow x \leq s.$$

This is an internal statement (with standard parameters A and s). By transfer, it then holds for *all* x, which means that s is an upper bound of A.

Now s is the *least* upper bound of A because suppose there is an upper bound t of A with $t < s$. By transfer, there would then be a standard t with this property. Since the difference between different standard numbers cannot be infinitesimal, $t < a + (m-1)\alpha$ follows. However, this would mean that t is not an upper bound of A, in contradiction to the assumption. □

In $\mathbb{R}$ there are infinitely many nonstandard numbers between two standard numbers. In $\mathbb{N}$, on the other hand, we expect every nonstandard number to be greater than all standard numbers or, conversely, that all numbers that are smaller than a standard number are also standard. This is confirmed by the following theorem.

Theorem 13. *Let $n \in \mathbb{N}$ be standard. Then all $m \in \mathbb{N}$ with $m < n$ are also standard.*

Proof. Let $n \in \mathbb{N}$ be standard and $m \in \mathbb{N}$, $m < n$. Then m is limited and, according to the standard part axiom, there is a standard $r \in \mathbb{R}$ with $m \approx r$. Hence $r - \frac{1}{2} < m < r + \frac{1}{2}$. Thus there is $k \in \mathbb{N}$ with $r - \frac{1}{2} < k < r + \frac{1}{2}$. This is an internal statement (with standard parameter r). Therefore, the transfer axiom is applicable and it follows that there is a standard $k \in \mathbb{N}$ with $r - \frac{1}{2} < k < r + \frac{1}{2}$. Since the open interval $]r - \frac{1}{2}, r + \frac{1}{2}[$ can contain at most one integer, $m = k$ follows. Hence m is standard.[12] □

2.7 Concluding remarks on the axiomatics

A major challenge in the application of the nonstandard theory is to avoid illegal transfer and illegal set formation. When applying the transfer axiom, it must always be ensured that the statement to be transferred is *internal* and that all its *parameters are standard*. When forming sets by separation, we must always make sure that the statement used for separation is *internal*. We will discuss the latter in more detail.

The fact that the axiom schema of separation only applies to internal statements does not imply any restriction compared to standard analysis, but it seems unsatisfactory when viewed with a naive understanding of sets. Why should it not be possible to separate from any set the subset of, for example, all its standard elements?

First of all, it should be noted that a naive understanding of sets based on comprehension with arbitrary predicates has led to contradictions and that, at the beginning of the twentieth century, an axiomatically founded concept of sets was therefore sought. Since then, axioms have regulated which fundamental statements apply to sets.

---

[12] The proof is taken from a draft version of [104]. In IST (with a much more general idealization axiom), the theorem follows directly from Theorem 52 in Section 3.5.3 (the characterization of finite standard sets).

In Zermelo–Fraenkel's set theory, the general axiom of comprehension is weakened to the axiom of separation. Instead of the intuitively plausible set formation according to the schema $\{x \mid \varphi(x)\}$, only the separation from already given sets is allowed, i. e., the set formation according to the schema $\{x \in A \mid \varphi(x)\}$ (whereby the existence of A must be ensured beforehand). What is the motivation for not allowing this separation schema for the new predicate *standard*?

We find ourselves in a situation similar to that at the end of the nineteenth century, when it was decided to allow actual infinite sets in mathematical considerations. To do this, we had to abandon a plausible principle that was thousands of years old and belonged to the Euclidean axioms: The whole is greater than its part. We have already mentioned that Leibniz rejected infinite wholes precisely because they violate Euclid's part-whole axiom.

If we want to have actual infinite sets, we have to sacrifice this axiom in order to avoid contradictions. And if we want to allow the existence of natural numbers beyond all standard numbers (which include all numbers definable in conventional analysis), then we have to sacrifice the separation with the predicate that distinguishes standard and nonstandard numbers.[13]

If a set formation like $\{x \in \mathbb{N} \mid x \text{ is standard}\}$ were possible, then it would contradict the (proven) well-ordering principle because the complement of this set in $\mathbb{N}$ would be nonempty, but would not have a smallest element.

How should we imagine the set $\mathbb{N}$? A sequence of standard numbers open at the end is followed by a sequence of nonstandard numbers open at the beginning (and at the end) (see Figure 2.2). Among the standard numbers there is no last one (with n also $n+1$ is standard), and among the nonstandard numbers there is no first one (with n also $n-1$ is nonstandard). However, it is not possible to separate the set $\mathbb{N}$ exactly between the standard and nonstandard numbers because this would be a separation with the external predicate *standard*. In contrast, it is possible to split the set $\mathbb{N}$ at any number (standard or nonstandard) without any problems.

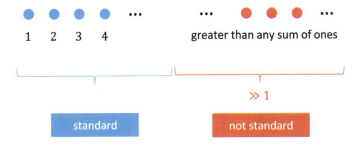

Figure 2.2: The set of natural numbers in the nonstandard view.

---

[13] In *external set theories* (see Section 3.6.2), separation with external statements is possible. However, other set axioms must be restricted for this.

A potentially helpful analogy from everyday life is the *sorites paradox* (also known as the *paradox of the heap*), which describes the difficulty of precisely defining a vague concept such as a heap (see [62], pp. 244–258). When is a collection of grains of sand a heap? Intuitively, there should be so many that the collection is still a heap after one grain of sand has been removed. On the other hand, a single grain of sand does not intuitively form a heap. It is not possible to specify an exact and intuitively plausible boundary between heaps and nonheaps.

Similarly, it is not possible to use the axioms of set theory to draw a boundary between standard and nonstandard numbers in the natural numbers or a boundary between limited and unlimited numbers in the real numbers (the latter would contradict the least-upper-bound property).

! Given the representation in Figure 2.2, one may be tempted to identify the standard numbers in $\mathbb{N}$ with the sums of ones. Such an idea may even be helpful for heuristic considerations. In fact, however, it is only certain that every sum of ones is standard and that every nonstandard number in $\mathbb{N}$ is greater than every sum of ones.[14]

2.8 External criteria for central concepts of analysis

In calculus courses the central concepts of analysis (convergence, continuity, differentiability, integrability, etc.) are usually defined conventionally, i. e., by *internal* statements. We assume that these definitions are known.

In the following sections, we specify logically simpler, *external* criteria that will prove to be equivalent to the logically more complex, internal definitions for standard parameters (e. g., in relation to standard functions and standard numbers).

The benefit of the external criteria is that an alternative method of proof is available for conventional (i. e., internally formulated) theorems according to the following schema:

1. Proof of the theorem for standard parameters (standard functions and standard numbers) using the external criteria.
2. Generalization of the statement by transfer (this is possible because the statement is internal and has only standard parameters).

Alternatively, it is also possible to *define* the central concepts of analysis by nonstandard criteria without assuming knowledge of the conventional definitions provided that the

---

14 Formally (and with the unary relation symbol st for *standard*) this could be formulated as a *theorem schema*: For each sum term τ consisting of ones, we have the theorem
$$\text{st}(\tau) \wedge \forall n \in \mathbb{N}\left(\neg\text{st}(n) \Rightarrow n > \tau\right).$$
A *sum term consisting of ones* can be defined recursively in the metalanguage. The theorem schema can then be proved via metalanguage induction on the number of summands in τ.

axiom schema of standardization is included in the axiomatics. We present this possibility in Section 2.10. However, we will want to clarify the relationship between the conventional definitions and the external criteria anyway. Therefore, the standardization schema is not mandatory (for basic analysis). The external criteria can then be used (as described above) to enrich the spectrum of methods.

In the following, let D always be a subset of $\mathbb{R}$. If $f: D \to \mathbb{R}$ is a function and standard, then its domain D is also standard.

To simplify logical formulas in the proofs, we also stipulate that the variables $a, b, c, x, y, \varepsilon, \delta$ should stand for real numbers and the variables m, n for natural numbers. Accordingly, for example, $\forall x\, \exists n \ldots$ stands for $\forall x \in \mathbb{R}\, \exists n \in \mathbb{N} \ldots$.

2.8.1 Convergence of sequences

Theorem 14. *Let $f: \mathbb{N} \to \mathbb{R}$ be a standard sequence. Then f is convergent if and only if there is a standard $a \in \mathbb{R}$ such that for all $n \gg 1$,*

$$f(n) \approx a.$$

In this case, a is the limit of f and

$$a = \lim_{k \to \infty} f(k) = \mathrm{stp}(f(n)) \quad \text{for all } n \gg 1.$$

Proof. According to the conventional definition, f is convergent if there is an a such that

$$\forall \varepsilon > 0\, \exists m\, \forall n\, (n \geq m \Rightarrow |f(n) - a| < \varepsilon). \tag{2.6}$$

We have to show that this is the case if and only if there is a standard a such that

$$\forall n\, (n \gg 1 \Rightarrow f(n) \approx a). \tag{2.7}$$

First, we show that (2.6) and (2.7) are equivalent for any standard a. So let a be standard. Then

$$\forall \varepsilon > 0\, \exists m\, \forall n\, (n \geq m \Rightarrow |f(n) - a| < \varepsilon)$$

$\overset{(1)}{\Leftrightarrow}$ $\forall^{\mathrm{st}} \varepsilon > 0\, \exists^{\mathrm{st}} m\, \forall n\, (n \geq m \Rightarrow |f(n) - a| < \varepsilon)$

$\overset{(2)}{\Leftrightarrow}$ $\forall^{\mathrm{st}} \varepsilon > 0\, \forall n\, (n \gg 1 \Rightarrow |f(n) - a| < \varepsilon)$

$\overset{(3)}{\Leftrightarrow}$ $\forall n\, \forall^{\mathrm{st}} \varepsilon > 0\, (n \gg 1 \Rightarrow |f(n) - a| < \varepsilon)$

$\overset{(4)}{\Leftrightarrow}$ $\forall n\, (n \gg 1 \Rightarrow \forall^{\mathrm{st}} \varepsilon > 0\, |f(n) - a| < \varepsilon)$

$\overset{(5)}{\Leftrightarrow}$ $\forall n\, (n \gg 1 \Rightarrow f(n) \approx a).$

Explanations. (1) By transfer (first applied to $\forall \varepsilon > 0 \dots$ and then in the dual variant to $\exists m \dots$).

(2) "$\Rightarrow$" If there is a standard m such that $|f(n) - a| < \varepsilon$ holds for all $n \geq m$, then $|f(n) - a| < \varepsilon$ holds a fortiori for all $n \gg 1$.

"$\Leftarrow$" If $|f(n) - a| < \varepsilon$ holds for all $n \gg 1$, then there is an m (for example, an arbitrary unlimited one) such that $\forall n\, (n \geq m \Rightarrow |f(n) - a| < \varepsilon)$. This is an internal statement about m (with standard parameters f, a, ε). By transfer, there is then also a standard m with this property.

(3) Because neighboring universal quantifiers can always be interchanged.

(4) Because the premise $n \gg 1$ does not depend on ε.

(5) According to the definition of "$\approx$".

If f is convergent, then a is uniquely determined by the internal statement (2.6) as $\lim_{k \to \infty} f(k)$, hence standard. Therefore a also satisfies (2.7), and we have $a = \lim_{k \to \infty} f(k) = \mathrm{stp}(f(n))$ for all $n \gg 1$.

Conversely, if there is a standard a with (2.7), then a also satisfies (2.6). □

Theorem 15. *Let $f: \mathbb{N} \to \mathbb{R}$ be a standard sequence. Then f is a Cauchy sequence if and only if for all $m, n \gg 1$,*

$$f(m) \approx f(n).$$

Proof. Exercise. □

Theorem 16. *Let $f: \mathbb{N} \to \mathbb{R}$ and $a \in \mathbb{R}$ both be standard. Then a is a cluster point of the sequence f if and only if there is an $n \gg 1$ with $f(n) \approx a$.*

Proof. By definition, a is a cluster point of f if every ε-neighborhood of a contains infinitely many sequence members. This is equivalent to

$$\forall \varepsilon > 0\; \forall m\; \exists n\, (n \geq m \wedge |f(n) - a| < \varepsilon)$$

and, due to the Archimedean axiom, also equivalent to

$$\forall m\, \exists n \left(n \geq m \wedge |f(n) - a| < \frac{1}{m} \right). \tag{2.8}$$

We have to show that this is equivalent to

$$\exists n\, (n \gg 1 \wedge f(n) \approx a). \tag{2.9}$$

This results from the following equivalence transformation:

$$\forall m\, \exists n \left(n \geq m \wedge |f(n) - a| < \frac{1}{m} \right)$$

$$\overset{(1)}{\Leftrightarrow} \quad \forall^{st} m \exists n \left(n \geq m \wedge |f(n) - a| < \frac{1}{m} \right)$$

$$\overset{(2)}{\Leftrightarrow} \quad \exists n \forall^{st} m \left(n \geq m \wedge |f(n) - a| < \frac{1}{m} \right)$$

$$\overset{(3)}{\Leftrightarrow} \quad \exists n \, (n \gg 1 \wedge f(n) \approx a).$$

Explanations. (1) By transfer.
(2) "$\Rightarrow$" Let $M := \{m \mid \exists n \, (n \geq m \wedge |f(n) - a| < \frac{1}{m})\}$. Then M contains all standard m by assumption. If M contained only standard numbers, then $\mathbb{N} \setminus M$ would be nonempty and contain a smallest number n_0 according to the well-ordering principle. Then n_0 would not be standard, but $n_0 - 1 \in M$ would be standard. Contradiction! Hence M must also contain nonstandard numbers. For such a nonstandard number $m_0 \in M$, the following holds (according to the definition of M): $\exists n \, (n \geq m_0 \wedge |f(n) - a| < \frac{1}{m_0})$. This means (since m_0 is greater than all standard m), there is an n such that $n \geq m \wedge |f(n) - a| < \frac{1}{m}$ holds for all standard m.
"$\Leftarrow$" Trivial.
(3) According to the definition of "$\gg$" and "$\approx$" in combination with Theorem 9. □

2.8.2 Continuity at a point

Theorem 17. *Let $f : D \to \mathbb{R}$ and $a \in D$ both be standard. Then f is continuous at a if and only if for all $x \in D$ with $x \approx a$,*

$$f(x) \approx f(a).$$

Proof. According to the ε-δ definition of continuity, f is continuous at a provided that

$$\forall \varepsilon > 0 \, \exists \delta > 0 \, \forall x \in D \, (|x - a| < \delta \Rightarrow |f(x) - f(a)| < \varepsilon). \tag{2.10}$$

We have to show that this is the case if and only if

$$\forall x \in D \, (x \approx a \Rightarrow f(x) \approx f(a)). \tag{2.11}$$

This results from the following equivalence transformation:

$$\forall \varepsilon > 0 \, \exists \delta > 0 \, \forall x \in D \, (|x - a| < \delta \Rightarrow |f(x) - f(a)| < \varepsilon)$$

$$\overset{(1)}{\Leftrightarrow} \quad \forall^{st} \varepsilon > 0 \, \exists^{st} \delta > 0 \, \forall x \in D \, (|x - a| < \delta \Rightarrow |f(x) - f(a)| < \varepsilon)$$

$$\overset{(2)}{\Leftrightarrow} \quad \forall^{st} \varepsilon > 0 \, \forall x \in D \, (x \approx a \Rightarrow |f(x) - f(a)| < \varepsilon)$$

$$\overset{(3)}{\Leftrightarrow} \quad \forall x \in D \, \forall^{st} \varepsilon > 0 \, (x \approx a \Rightarrow |f(x) - f(a)| < \varepsilon)$$

$$\overset{(4)}{\Leftrightarrow} \quad \forall x \in D \, (x \approx a \Rightarrow \forall^{st} \varepsilon > 0 \, |f(x) - f(a)| < \varepsilon)$$

$$\overset{(5)}{\Leftrightarrow} \quad \forall x \in D \, (x \approx a \Rightarrow f(x) \approx f(a)).$$

Explanations. (1) By transfer.
(2) "$\Rightarrow$" If $|x - a| < \delta$ (with a standard $\delta > 0$) is sufficient for $|f(x) - f(a)| < \varepsilon$, then in particular $x \approx a$ is sufficient.
"$\Leftarrow$" If $\forall x \in D \, (x \approx a \Rightarrow |f(x) - f(a)| < \varepsilon)$ holds, then there is a $\delta > 0$ (for example, an arbitrary infinitesimal one) such that

$$\forall x \in D \, (|x - a| < \delta \Rightarrow |f(x) - f(a)| < \varepsilon). \tag{2.12}$$

This is an internal statement about δ (with standard parameters f, D, a, ε). By transfer, there is then also a standard $\delta > 0$ that satisfies (2.12).
(3) Because neighboring universal quantifiers can always be interchanged.
(4) Because the premise $x \approx a$ does not depend on ε.
(5) According to the definition of "$\approx$". □

2.8.3 Uniform continuity

Theorem 18. *Let $f \colon D \to \mathbb{R}$ be standard. Then f is uniformly continuous if and only if for all $x_1, x_2 \in D$ with $x_1 \approx x_2$,*

$$f(x_1) \approx f(x_2).$$

Proof. By conventional definition, f is uniformly continuous provided that

$$\forall \varepsilon > 0 \, \exists \delta > 0 \, \forall x_1, x_2 \in D \, (|x_1 - x_2| < \delta \Rightarrow |f(x_1) - f(x_2)| < \varepsilon). \tag{2.13}$$

We have to show that this is the case if and only if

$$\forall x_1, x_2 \in D \, (x_1 \approx x_2 \Rightarrow f(x_1) \approx f(x_2)). \tag{2.14}$$

The equivalence transformation is analogous to that in the proof of Theorem 17 (exercise). □

2.8.4 Accumulation point of a set

Theorem 19. *Let $D \subseteq \mathbb{R}$ and $a \in D$ both be standard. Then a is an accumulation point of D if and only if there is an $x \in D \setminus \{a\}$ with $x \approx a$.*

Proof. According to the conventional definition, a is an accumulation point of D provided that

$$\forall \varepsilon > 0 \; \exists x \in D \setminus \{a\} \; |x - a| < \varepsilon. \tag{2.15}$$

We have to show that this is the case if and only if

$$\exists x \in D \setminus \{a\} \; x \approx a. \tag{2.16}$$

Statement (2.16) immediately follows from (2.15) with an arbitrary infinitesimal $\varepsilon > 0$. The reverse is also true, because (2.16) means

$$\exists x \in D \setminus \{a\} \; \forall^{st} \varepsilon > 0 \; |x - a| < \varepsilon.$$

This implies

$$\forall^{st} \varepsilon > 0 \; \exists x \in D \setminus \{a\} \; |x - a| < \varepsilon.$$

The part after $\forall^{st} \varepsilon > 0$ is an internal statement with standard parameters ε, D, and a. Therefore, transfer is applicable and (2.15) follows. □

2.8.5 Limits of functions

For $f: D \to \mathbb{R}$ and $a, b \in \mathbb{R}$, let $f_{(a,b)}: D \cup \{a\} \to \mathbb{R}$ be the function defined by $f_{(a,b)}(a) := b$ and $f_{(a,b)}(x) := f(x)$ for $x \neq a$. If f, a, b are standard, then the function $f_{(a,b)}$ (uniquely determined by an internal statement with standard parameters) is also standard.

Theorem 20. *Let $f: D \to \mathbb{R}$ be standard and let a be a standard accumulation point of D. Then $f(x)$ is convergent as $x \to a$ if and only if there is a standard $b \in \mathbb{R}$ such that for all $x \in D \setminus \{a\}$ with $x \approx a$,*

$$f(x) \approx b.$$

In this case, b is the limit *of f (as x approaches a) and*

$$b = \lim_{x \to a} f(x) = \operatorname{stp}(f(x)), \quad \text{for all } x \in D \setminus \{a\} \text{ with } x \approx a.$$

Proof. According to the ε–δ definition for limits of functions, $f(x)$ is convergent as $x \to a$ if and only if there is a b such that

$$\forall \varepsilon > 0 \; \exists \delta > 0 \; \forall x \in D \setminus \{a\} \; (|x - a| < \delta \Rightarrow |f(x) - b| < \varepsilon), \tag{2.17}$$

i.e., if and only if there is a b such that $f_{(a,b)}$ is continuous in a.

We have to show that this is the case if and only if there is a standard b such that

$$\forall x \in D \setminus \{a\} \; (x \approx a \Rightarrow f(x) \approx b). \tag{2.18}$$

If $f(x)$ is convergent as $x \to a$, then $b = \lim_{x \to a} f(x)$ is uniquely determined by (2.17) and is hence standard. Therefore, $f_{(a,b)}$ is also standard and, by definition and considering (2.17), continuous at a. By Theorem 17, then

$$\forall x \in D \, (x \approx a \Rightarrow f_{(a,b)}(x) \approx f_{(a,b)}(a)) \tag{2.19}$$

and hence (2.18), as well as $b = \lim_{x \to a} f(x) = \mathrm{stp}(f(x))$ for all $x \in D \setminus \{a\}$ with $x \approx a$.

Conversely, if there is a standard b with (2.18), then (2.19) follows for such a b. Thus $f_{(a,b)}$ is continuous at a according to Theorem 17 and hence f is convergent as $x \to a$. □

2.8.6 Differentiability

Theorem 21. *Let $f : D \to \mathbb{R}$ be standard and let $a \in D$ be a standard accumulation point of D. Then f is differentiable at a if and only if there is a standard $b \in \mathbb{R}$ such that for all $x \in D \setminus \{a\}$ with $x \approx a$,*

$$\frac{f(x) - f(a)}{x - a} \approx b.$$

In this case, b is the derivative *of f at a and*

$$b = f'(a) = \mathrm{stp}\left(\frac{f(x) - f(a)}{x - a}\right), \quad \text{for all } x \in D \setminus \{a\} \text{ with } x \approx a.$$

Proof. Let $g : D \setminus \{a\} \to \mathbb{R}$ be defined by

$$g(x) := \frac{f(x) - f(a)}{x - a}.$$

Since f and a are standard, g is also standard.

By conventional definition, f is differentiable at a if and only if g is convergent as $x \to a$. By Theorem 20, this is the case if and only if there is a standard b as required by Theorem 21. □

Remarks. 1. In the conventional definition of differentiability at a point a, it is assumed that a is an accumulation point of D. This is also assumed in Theorem 21 and ensures (according to Theorem 19) that there is an $x \in D \setminus \{a\}$ with $x \approx a$.
2. As an immediate conclusion from Theorem 21, we obtain: If f and a are standard and f is differentiable at a, then for all $x \approx a$,

$$f(x) = f(a) + f'(a)(x - a) + o(x, a), \tag{2.20}$$

with $\frac{o(x,a)}{x-a} \approx 0$.

2.8.7 Integrability

In conventional analysis, the integral of a function can be defined by Riemann sums, which are nothing other than integrals of step functions (i. e., finite sums) that interpolate the integrand function at certain intermediate points. The integral exists if the Riemann sums are arbitrarily close to a fixed value (independent of the partition of the interval and the choice of intermediate points), provided the partition is sufficiently fine.

This definition can be translated into a corresponding nonstandard description. The integral is then the common standard part of the Riemann sums for infinitely fine partitions (provided that such a common standard part exists).

Definition 5. Let $a, b \in \mathbb{R}$ and $a < b$. A finite strictly monotonically increasing sequence $(x_i)_{0 \leq i \leq n}$ with $x_0 = a$ and $x_n = b$ is called a *finite partition of* $[a, b]$.

The positive real number $\delta := \max\{x_i - x_{i-1} \mid 1 \leq i \leq n\}$ is called the *mesh* of the partition. The partition is called *infinitely fine* if $\delta \approx 0$.

Furthermore, if a sequence $(\xi_i)_{1 \leq i \leq n}$ of intermediate points $\xi_i \in [x_{i-1}, x_i]$ is given, then the pair $P := ((x_i)_{0 \leq i \leq n}, (\xi_i)_{1 \leq i \leq n})$ is called a *Riemann partition of* $[a, b]$ *(with division points x_i and intermediate points ξ_i).*[15] The mesh of $(x_i)_{0 \leq i \leq n}$ is called the mesh of P.

Definition 6. Let $f : [a, b] \to \mathbb{R}$ and $P := ((x_i)_{0 \leq i \leq n}, (\xi_i)_{1 \leq i \leq n})$ a Riemann partition of $[a, b]$. Then

$$R(P, f) := \sum_{i=1}^{n} f(\xi_i)(x_i - x_{i-1})$$

is called the *Riemann sum of P and f*.

The function $f_P : [a, b] \to \mathbb{R}$ with $f_P(x) = f(\xi_i)$ for $x \in [x_{i-1}, x_i[, 1 \leq i \leq n$ and $f_P(b) = f(\xi_n)$ is called *the step function of f associated with P*.

Due to the definition of the integral for step functions,

$$\int_a^b f_P(x)\, dx = \sum_{i=1}^{n} f(\xi_i)(x_i - x_{i-1}) = R(P, f). \qquad (2.21)$$

Theorem 22. *Let $f : D \to \mathbb{R}$ be standard and $a, b \in \mathbb{R}$ also be standard. Moreover, let $a < b$ and $[a, b] \subseteq D$. Then f is integrable from a to b if and only if there is a standard $c \in \mathbb{R}$ such that for every infinitely fine Riemann partition P of $[a, b]$,*

$$R(P, f) \approx c. \qquad (2.22)$$

In this case, c is the integral of f (from a to b) and

---

[15] The term *Riemann partition* is not commonly used in the literature, but simplifies many formulations in the following.

$$c = \int_a^b f(x)\,dx = \operatorname{stp}(R(P,f))$$

for all infinitely fine Riemann partitions P of $[a,b]$.

Proof. Let $\varphi(P)$ be the statement "P is a Riemann partition of $[a,b]$" and F be the function that assigns to each Riemann partition of $[a,b]$ its mesh. Both φ and F depend on the standard parameters a and b, which are suppressed in the designation for the sake of brevity.

According to Riemann's integral definition, f is integrable from a to b provided there is a $c \in \mathbb{R}$ such that

$$\forall \varepsilon > 0\ \exists \delta > 0\ \forall P\,(\varphi(P) \wedge F(P) < \delta \Rightarrow |R(P,f) - c| < \varepsilon). \tag{2.23}$$

We have to show that this is the case if and only if there is a standard $c \in \mathbb{R}$ such that

$$\forall P\,(\varphi(P) \wedge F(P) \approx 0 \Rightarrow R(P,f) \approx c). \tag{2.24}$$

This results from the following equivalence transformation:

$$\forall \varepsilon > 0\ \exists \delta > 0\ \forall P\,(\varphi(P) \wedge F(P) < \delta \Rightarrow |R(P,f) - c| < \varepsilon)$$

$\overset{(1)}{\Leftrightarrow}\ \forall^{st} \varepsilon > 0\ \exists^{st} \delta > 0\ \forall P\,(\varphi(P) \wedge F(P) < \delta \Rightarrow |R(P,f) - c| < \varepsilon)$

$\overset{(2)}{\Leftrightarrow}\ \forall^{st} \varepsilon > 0\ \forall P\,(\varphi(P) \wedge F(P) \approx 0 \Rightarrow |R(P,f) - c| < \varepsilon)$

$\overset{(3)}{\Leftrightarrow}\ \forall P\, \forall^{st} \varepsilon > 0\,(\varphi(P) \wedge F(P) \approx 0 \Rightarrow |R(P,f) - c| < \varepsilon)$

$\overset{(4)}{\Leftrightarrow}\ \forall P\,(\varphi(P) \wedge F(P) \approx 0 \Rightarrow \forall^{st} \varepsilon > 0\ |R(P,f) - c| < \varepsilon)$

$\overset{(5)}{\Leftrightarrow}\ \forall P\,(\varphi(P) \wedge F(P) \approx 0 \Rightarrow R(P,f) \approx c).$

Explanations. (1) By transfer.

(2) "$\Rightarrow$" If $\varphi(P) \wedge F(P) < \delta$ (with a standard $\delta > 0$) is sufficient for

$$|R(P,f) - c| < \varepsilon,$$

then in particular $\varphi(P) \wedge F(P) \approx 0$ is sufficient.

"$\Leftarrow$" If $\forall P\,(\varphi(P) \wedge F(P) \approx 0 \Rightarrow |R(P,f) - c| < \varepsilon)$ holds, then there is a $\delta > 0$ (for example, an arbitrary infinitesimal one) such that

$$\forall P\,(\varphi(P) \wedge F(P) < \delta \Rightarrow |R(P,f) - c| < \varepsilon) \tag{2.25}$$

This is an internal statement about δ with standard parameters f, a, b, c, ε. By transfer, there is then also a standard $\delta > 0$ that satisfies (2.25).

(3) Because neighboring universal quantifiers can always be interchanged.
(4) Because the premise $\varphi(P) \wedge F(P) \approx 0$ does not depend on ε.
(5) According to the definition of "$\approx$". □

Note that the equivalence transformations in the proofs of the Theorems 14, 17, and 22 are completely analogous.

2.9 Some theorems and proofs

2.9.1 Theorems about sequences

As usual, we now write sequences in the form $(a_n)_{n \in \mathbb{N}}$ or simply (a_n), which means that n is mapped to a_n for all $n \in \mathbb{N}$.

Theorem 23. *Every Cauchy sequence converges.*

Proof. Let (a_n) initially be standard and let $a_m \approx a_n$ for all $m, n \gg 1$. Then (a_n) is bounded, that is, there is an $r \in \mathbb{R}$ with $|a_n| \leq r$ for all $n \in \mathbb{N}$ (otherwise there would be an $m > n$ with $|a_m| > |a_n| + 1$ for all n in contradiction to $a_m \approx a_n$ for all $m, n \gg 1$). Since (a_n) is standard, it is then (by transfer) also bounded by a standard $r \in \mathbb{R}$. Hence all a_n are limited. Choose $m \gg 1$ and set $a := \text{stp}(a_m)$. Then $a_n \approx a_m \approx a$ holds for all $n \gg 1$. This means that (a_n) converges to a.

For general sequences, the assertion follows by transfer. □

Theorem 24. *Let (a_n) and (b_n) be sequences and let (a_n) converge to a and (b_n) converge to b. Then*
1. *$(a_n + b_n)$ converges to $a + b$,*
2. *$(a_n \cdot b_n)$ converges to $a \cdot b$,*
3. *if $b \neq 0$ and $b_n \neq 0$ for all $n \in \mathbb{N}$, then $(\frac{a_n}{b_n})$ converges to $\frac{a}{b}$,*
4. *if $a_n \leq b_n$ for all $n \in \mathbb{N}$, then $a \leq b$.*

Proof. For standard sequences, the assertion follows from Theorems 11 and 14. The general case follows by transfer. □

The calculation rules for limits do not play a major role in nonstandard analysis, as calculations are usually performed directly with the standard parts.

2.9.2 Theorems about continuous functions

Theorem 25. *Let $f, g : D \to \mathbb{R}$ be continuous, $c \in \mathbb{R}$, and $D' = \{x \in D \mid g(x) \neq 0\}$. Then the functions $f + g, cf, fg : D \to \mathbb{R}$, and $\frac{f}{g} : D' \to \mathbb{R}$ are continuous.*

Proof. For standard parameters, this follows directly from the calculation rules for infinitesimal numbers and generally by transfer. □

Theorem 26. *Let $f: D \to \mathbb{R}$, $g: E \to \mathbb{R}$, and $f(D) \subseteq E$. Further, let f be continuous at $a \in D$ and g continuous at $b := f(a)$. Then the function $g \circ f: D \to \mathbb{R}$ is continuous at a.*

Proof. Let f, g and a initially be standard. Then $b := f(a)$ is also standard. Now let $x \in D$, $x \approx a$. Since f is continuous at a, $f(x) \approx b$ follows. Since g is continuous at b, $g(f(x)) \approx g(b) = g(f(a))$ follows. Hence $g \circ f$ is continuous at a.

For general f, g, a the assertion follows by transfer. □

Theorem 27 (Bolzano's theorem). *Let $f: [a,b] \to \mathbb{R}$ be continuous, $f(a) < 0$, and $f(b) > 0$. Then there is a $c \in [a,b]$ with $f(c) = 0$.*

Proof. Let f, a, b initially be standard. Choose an arbitrary $n \gg 1$ and set $x_i := a + i \cdot \frac{b-a}{n}$ for $i = 0, \ldots, n$. Let k be the smallest of the indices i with $f(x_i) \geq 0$. Then

$$1 \leq k \leq n \quad \text{and} \quad f(x_{k-1}) < 0 \quad \text{and} \quad f(x_k) \geq 0. \tag{2.26}$$

Because of $n \gg 1$, we have $x_{k-1} \approx x_k$. Since a, b are standard and $a \leq x_k \leq b$, the standard part $c := \text{stp}(x_k)$ exists, and (by Theorem 11, point 4)

$$a = \text{stp}(a) \leq c \leq \text{stp}(b) = b.$$

From $x_{k-1} \approx c \approx x_k$ and the continuity of f follows $f(x_{k-1}) \approx f(c) \approx f(x_k)$, and thus

$$|f(x_k)| \leq |f(x_k)| + |f(x_{k-1})| = f(x_k) - f(x_{k-1}) \approx 0,$$

hence $f(c) \approx f(x_k) \approx 0$. With f and c, $f(c)$ is also standard and $f(c) = 0$ follows.

For general f, a, b, the assertion follows by transfer. □

As usual, Bolzano's theorem can be generalized to the intermediate value theorem.

Theorem 28. *Let $f: [a,b] \to \mathbb{R}$ be continuous. Then f is uniformly continuous.*

Proof. Let f, a, b initially be standard. For all x with $a \leq x \leq b$,

$$a = \text{stp}(a) \leq \text{stp}(x) \leq \text{stp}(b) = b \tag{2.27}$$

follows from Theorem 11 (and because a and b are standard). Therefore, for all $x_1, x_2 \in [a,b]$ the respective standard parts are also in $[a,b]$. If $x_1 \approx x_2$, their standard parts are equal. Since f is continuous, it follows that

$$f(x_1) \approx f(\text{stp}(x_1)) = f(\text{stp}(x_2)) \approx f(x_2).$$

This means that f is uniformly continuous, according to Theorem 18.

For general f, a, b, the assertion follows by transfer. □

Theorem 29. *Let $f: [a,b] \to \mathbb{R}$ be continuous. Then f attains a minimum and a maximum on $[a,b]$.*

Proof. Let f, a, b initially be standard. Choose $n \gg 1$ and set $x_i := a + i \cdot \frac{b-a}{n}$ for $i = 0, \ldots, n$. Among the finitely many real numbers $f(x_i)$, let $f(x_{i_1})$ be the smallest and $f(x_{i_2})$ the largest. Then for all i,

$$f(x_{i_1}) \le f(x_i) \le f(x_{i_2}). \tag{2.28}$$

Since a, b are standard, x_{i_1}, x_{i_2} are limited and the standard parts $\check{x} := \mathrm{stp}(x_{i_1})$ and $\hat{x} := \mathrm{stp}(x_{i_2})$ exist. As with (2.27), $\check{x}, \hat{x} \in [a, b]$. Because f is continuous and standard, $f(\check{x}) = \mathrm{stp}(f(x_{i_1}))$ and $f(\hat{x}) = \mathrm{stp}(f(x_{i_2}))$.

For all standard $x \in [a, b]$, there is an i with $x = \mathrm{stp}(x_i)$ and (again because f is continuous and standard) $f(x) = \mathrm{stp}(f(x_i))$. Taking the standard parts in (2.28) yields

$$f(\check{x}) \le f(x) \le f(\hat{x}).$$

By transfer, it follows that this inequality holds for all $x \in [a, b]$. Therefore f attains its minimum at $\check{x}$ and its maximum at $\hat{x}$.

For general f, a, b, the assertion follows by transfer. □

2.9.3 Theorems of differential calculus

Theorem 30. *Let $f : D \to \mathbb{R}$ be differentiable at a. Then f is continuous at a.*

Proof. Let f, a initially be standard. If f is differentiable at a, then $f'(a)$ is standard (i. e., also limited) and, according to (2.20), for all $x \approx a$,

$$f(x) = f(a) + f'(a)(x - a) + o(x, a),$$

with $\frac{o(x,a)}{x-a} \approx 0$. Since $f'(a)$ is limited, $f(x) \approx f(a)$ follows. Hence f is continuous at a.

For general f, a the assertion follows by transfer. □

Theorem 31. *Let $f, g : D \to \mathbb{R}$ be differentiable and let $c \in \mathbb{R}$. Then the functions $f + g, cf, fg : D \to \mathbb{R}$ are differentiable and for all $x \in D$,*
1. $(f + g)'(x) = f'(x) + g'(x),$
2. $(cf)'(x) = cf'(x),$
3. $(fg)'(x) = f'(x)g(x) + f(x)g'(x).$

If $g(x) \ne 0$ for all $x \in D$, then $\frac{f}{g}$ is also differentiable on D and for all $x \in D$,

$$\left(\frac{f}{g}\right)'(x) = \frac{f'(x)g(x) - f(x)g'(x)}{g(x)^2}.$$

Proof. The differentiation rules can be calculated using the differentials $df = f(x+dx) - f(x)$ and $dg = g(x+dx) - g(x)$. As an example, we calculate the product rule (point 3).

Let f, g, x initially be standard, $x \in D$ and $dx \approx 0$, $dx \neq 0$ with $x + dx \in D$. Then

$$d(fg) = f(x+dx)g(x+dx) - f(x)g(x)$$
$$= (f(x) + df)(g(x) + dg) - f(x)g(x)$$
$$= df \cdot g(x) + f(x) \cdot dg + df \cdot dg.$$

Division by dx yields

$$\frac{d(fg)}{dx} = \frac{df}{dx} \cdot g(x) + f(x) \cdot \frac{dg}{dx} + \frac{df}{dx} \cdot dg.$$

Since f, g, x are standard and f and g are differentiable at x, all terms are limited, and we can take the respective standard parts. Since g is continuous at x, $dg \approx 0$ follows, and we obtain $(fg)'(x) = f'(x)g(x) + f(x)g'(x)$.

For general f, g, x, the assertion follows by transfer. □

Theorem 32 (chain rule). *Let $f : V \to \mathbb{R}$ and $g : W \to \mathbb{R}$ be functions with $f(V) \subseteq W$. Further, let f be differentiable at $x \in V$ and g differentiable at $y := f(x) \in W$. Then the composed function $g \circ f : V \to \mathbb{R}$ is differentiable at x and*

$$(g \circ f)'(x) = g'(f(x))f'(x). \tag{2.29}$$

Proof. Let f, x initially be standard. Let $dx \approx 0$, $dx \neq 0$. We calculate with the differentials $df = f(x + dx) - f(x)$ and $dg = g(f(x) + df) - g(f(x))$. Since f is continuous at x, $df \approx 0$ follows. We have

$$d(g \circ f) = g(f(x+dx)) - g(f(x))$$
$$= g(f(x) + df) - g(f(x))$$
$$= g(f(x)) + dg - g(f(x))$$
$$= dg.$$

If $df = 0$, then $d(g \circ f) = 0$ already follows in the second line of the equation, i. e., $(g \circ f)'(x) = f'(x) = 0$ and thus (2.29). If $df \neq 0$, we have

$$\frac{d(g \circ f)}{dx} = \frac{dg}{dx} = \frac{dg}{df} \cdot \frac{df}{dx}.$$

Since f is differentiable at x and g is differentiable at $f(x)$, all terms are limited, and we can take their standard parts, which leads to (2.29).

For general f, x, the assertion follows by transfer. □

Theorem 33. *Let $f :]a, b[\to \mathbb{R}$ and $c \in]a, b[$. If f is differentiable at c and has a local extremum at c, then $f'(c) = 0$.*

Proof. Let f, c initially be standard. Let f have a local maximum in c (the case of a minimum is analogous). Since f and c are standard, by transfer there is a standard $\varepsilon > 0$

with $f(x) \le f(c)$ for all $x \in]c - \varepsilon, c + \varepsilon[$. For any positive $h \approx 0$, $f(c \pm h) - f(c) \le 0$ holds and therefore

$$\frac{f(c+h)-f(c)}{h} \le 0 \le \frac{f(c-h)-f(c)}{-h}.$$

By taking standard parts (and due to the differentiability of f at c),

$$f'(c) \le 0 \le f'(c),$$

hence $f'(c) = 0$.

For general f, c, the assertion follows by transfer. □

As usual, from Theorem 33 follow Rolle's theorem, the mean value theorem of differential calculus, and that functions which are continuous on $[a, b]$ and differentiable on $]a, b[$ with a vanishing derivative are constant.

2.9.4 Theorems of integral calculus

Theorem 34. *Let $f: [a, c] \to \mathbb{R}$ be integrable. Then for each $b \in [a, c]$ the restrictions of f on $[a, b]$ and $[b, c]$ are integrable and*

$$\int_a^c f(x)\,dx = \int_a^b f(x)\,dx + \int_b^c f(x)\,dx.$$

Proof. Follows directly from Theorem 22 for standard parameters and generally by transfer. □

Theorem 35. *Let $f, g: [a, b] \to \mathbb{R}$ be integrable and $f \le g$. Then*

$$\int_a^b f(x)\,dx \le \int_a^b g(x)\,dx.$$

Proof. Follows directly from Theorems 11 and 22 for standard parameters and generally by transfer. □

To show that continuous functions are integrable (Theorem 36), we need the following proposition.

Proposition 1. *Let $a, b \in \mathbb{R}$, $a < b$ and let $b - a$ be limited. Let f be a step function on $[a, b]$ and let $f(x) \approx 0$ for all $x \in [a, b]$. Then*

$$\int_a^b f(x)\,dx \approx 0.$$

Proof. Let $(x_i)_{0 \le i \le n}$ be a finite partition of $[a,b]$ and let $f(x) = c_i$ for $x \in \,]x_{i-1}, x_i[$, $1 \le i \le n$. Let $c := \max_{1 \le i \le n} |c_i|$. Then $c \approx 0$, and hence

$$\left| \int_a^b f(x)\,dx \right| = \left| \sum_{i=1}^n c_i(x_i - x_{i-1}) \right| \le \sum_{i=1}^n |c_i|(x_i - x_{i-1}) \le c(b-a) \approx 0. \qquad \square$$

Theorem 36. *Every continuous function $f : [a,b] \to \mathbb{R}$ is integrable.*

Proof. Let f, a, b initially be standard. We have to show that for every infinitely fine Riemann partition P the corresponding Riemann sum $R(P, f)$ is limited and that for every two infinitely fine Riemann partitions P, P' the Riemann sums differ only infinitesimally.

Let $P = ((x_i)_{0 \le i \le n}, (\xi_i)_{1 \le i \le n})$ be an infinitely fine Riemann partition of $[a,b]$. As a continuous function, f attains its minimum $\check{y}$ and its maximum $\hat{y}$ on $[a,b]$. Since f is standard, $\check{y}$ and $\hat{y}$ are also standard, and hence

$$\check{y}(b-a) = \check{y} \sum_{i=1}^n (x_i - x_{i-1}) \le \sum_{i=1}^n f(\xi_i)(x_i - x_{i-1}) \le \hat{y} \sum_{i=1}^n (x_i - x_{i-1}) = \hat{y}(b-a).$$

Hence $R(P, f)$ is limited.

Let $P = ((x_i)_{0 \le i \le n}, (\xi_i)_{1 \le i \le n})$ and $P' = ((x'_i)_{0 \le i \le n'}, (\xi'_i)_{1 \le i \le n'})$ be two infinitely fine Riemann partitions of $[a,b]$ and f_P resp. $f_{P'}$ the corresponding step functions according to Definition 6.

For all $x \in [a,b[$, there is an $i \in \{1, \ldots, n\}$ and a $j \in \{1, \ldots, n'\}$ with $x \in [x_{i-1}, x_i[$ and $x \in [x'_{j-1}, x'_j[$. Then $f_P(x) = f(\xi_i)$ with $\xi_i \in [x_{i-1}, x_i]$ and $f_{P'}(x) = f(\xi'_j)$ with $\xi'_j \in [x'_{j-1}, x'_j]$. For $x = b$, this is also true with $i = n$ and $j = n'$. Since P and P' are infinitely fine, $\xi_i \approx x \approx \xi'_j$. Since f is uniformly continuous on $[a,b]$, it follows that

$$f_P(x) = f(\xi_i) \approx f(x) \approx f(\xi'_j) = f_{P'}(x).$$

Thus $(f_P - f_{P'})(x) \approx 0$ for all $x \in [a,b]$. With (2.21) and Proposition 1 follows

$$R(P, f) - R(P', f) = \int_a^b f_P(x)\,dx - \int_a^b f_{P'}(x)\,dx = \int_a^b (f_P - f_{P'})(x)\,dx \approx 0.$$

For general f, a, b, the assertion follows by transfer. $\qquad \square$

From Theorem 36, together with Theorems 29 and 35, we obtain the following

Theorem 37. *Let $f : [a,b] \to \mathbb{R}$ be continuous and let $\check{y}, \hat{y}$ be its minimum and maximum on $[a,b]$, respectively. Then*

$$\check{y} \cdot (b-a) \le \int_a^b f(x)\,dx \le \hat{y} \cdot (b-a).$$

This theorem is also called the *rectangle property* of the integral, because the area under the curve described by f (i. e., the integral of f) is bounded below and above by the areas of two rectangles of width $b - a$ and height $y̌$ and $ŷ$, respectively.

2.9.5 Fundamental theorem of calculus

In the following, let $I \subseteq \mathbb{R}$ be a (bounded or unbounded, but nondegenerate) interval.

Theorem 38. *Let $f : I \to \mathbb{R}$ be continuous and $a \in I$. For each $x \in I$, let*

$$F(x) := \int_a^x f(t)\,dt.$$

Then $F : I \to \mathbb{R}$ is differentiable and $F' = f$.

Proof. We have to show that for any $x \in I$, F is differentiable at x and $F'(x) = f(x)$.

Let f, a, x initially be standard. If x is an interior point of I, there is an $\varepsilon > 0$ such that $]x - \varepsilon, x + \varepsilon[\subseteq I$. If x is a boundary point of I, there is an $\varepsilon > 0$ such that either $]x - \varepsilon, x] \subseteq I$ or $[x, x + \varepsilon[\subseteq I$. By transfer, there is then also a standard ε with the respective property in each case. Thus I also contains $x + dx$ for all positive infinitesimals dx or for all negative infinitesimals dx (or both).

Let $dx \approx 0$, $dx \neq 0$, and $x + dx \in I$. We assume $dx > 0$ (the case $dx < 0$ is similar). Define the differential

$$dF := F(x + dx) - F(x) = \int_x^{x+dx} f(t)\,dt.$$

In visual terms, dF is the increase in area under the curve when the right-hand boundary moves from x to $x + dx$. According to Theorem 37, we have

$$y̌ \cdot dx \leq dF \leq ŷ \cdot dx, \tag{2.30}$$

where $y̌$ and $ŷ$ are the minimum and maximum of f on $[x, x + dx]$, respectively (see Figure 2.3). Since f is uniformly continuous on the closed infinitesimal interval $[x, x+dx]$, $y̌$ and $ŷ$ are infinitely close to the standard number $f(x)$ which is therefore their common standard part. Division by dx and taking standard parts in (2.30) yields

$$f(x) = \mathrm{stp}(y̌) \leq \mathrm{stp}\left(\frac{dF}{dx}\right) \leq \mathrm{stp}(ŷ) = f(x). \tag{2.31}$$

Hence F is differentiable at x and $F'(x) = f(x)$.

For general f, a, x, the assertion follows by transfer. □

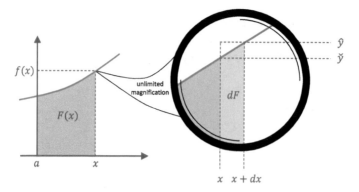

Figure 2.3: Nonstandard proof of the fundamental theorem.

Theorem 39. *Let $f: I \to \mathbb{R}$ be continuous and let F be an antiderivative of f (i. e., $F' = f$). Then for all $a, b \in I$,*

$$\int_a^b f(x)\, dx = F(b) - F(a).$$

Proof. This follows as usual from Theorem 38 because the difference between two antiderivatives is constant. □

2.10 Nonstandard definitions

In this chapter we have assumed the conventional, internal definitions of the central concepts of basic analysis to be known, and in Section 2.8 we have given external criteria that are equivalent to the conventional definitions, provided that their parameters are standard. The question arises whether these concepts can be *defined* by the external criteria without knowing about the existence of equivalent internal definitions. This is indeed possible if one has the standardization axiom S of internal set theory or the weakened, "parameter-free" version of S available (see Section 2.10.1 or 2.11.3).

2.10.1 Axiom schema of standardization

By transfer, every set that is defined by an internal statement is standard. As we know, standard sets can also contain nonstandard elements. According to Theorem 7, however, every standard set is already uniquely determined by its standard elements. It therefore seems quite reasonable to assume that one can *define* a standard set by specifying a (possibly external) predicate that determines its standard elements (knowing full well that this kind of definition might "automatically" add nonstandard elements). As a simple example, consider the standard elements of an interval. Let $\varphi(x)$ be the following statement:

$$x \text{ is standard, } x \geq 0, \text{ and } x \leq 1. \tag{2.32}$$

Since $\varphi(x)$ is external, $\{x \in \mathbb{R} \mid \varphi(x)\}$ would be an *illegal set formation*, i. e., not covered by the axiom schema of separation. However, there is a *standard set* (namely, the real interval $[0,1]$) whose standard elements are precisely those $x \in \mathbb{R}$ that satisfy $\varphi(x)$.

The following axiom schema is introduced in order to allow this type of set formation with external statements in general.

Axiom schema (standardization S). *Let $\varphi(x)$ be an (internal or external) statement (possibly with parameters). Then for every standard set A there is a standard set B whose standard elements are precisely those standard elements of A that satisfy $\varphi(x)$.*

According to Theorem 7, the standard set B is uniquely determined by A and φ and is a subset of A. It is denoted by

$$^S\{x \in A \mid \varphi(x)\}.$$

In a way, it can be seen as a "legal offer" instead of the set formation $\{x \in A \mid \varphi(x)\}$, which would be illegal if $\varphi(x)$ is external. If $\varphi(x)$ is internal and all its parameters are standard, both sets are identical (by transfer).

Note that the separation schema and the standardization schema have a completely analogous structure. The only differences are that in the standardization schema all quantifiers are relativized by the predicate *standard* and the statement $\varphi(x)$ may be external. While the extension of the separation schema to external statements would immediately lead to contradictions (as we have seen in Section 2.7), the (relative) consistency is maintained when the standardization schema is added (see Section 2.11.3).

With $\varphi(x)$ from (2.32), we can now write

$$^S\{x \in \mathbb{R} \mid \varphi(x)\} = [0,1]$$

because the standard $x \in \mathbb{R}$ with $\varphi(x)$ are exactly the standard $x \in [0,1]$. Some more examples of standardized sets:

$$^S\{n \in \mathbb{N} \mid n \text{ is standard}\} = \mathbb{N}, \tag{2.33}$$

$$^S\{n \in \mathbb{N} \mid n \text{ is not standard}\} = \emptyset, \tag{2.34}$$

$$^S\{x \in \mathbb{R} \mid x \text{ is limited}\} = \mathbb{R}, \tag{2.35}$$

$$^S\{x \in \mathbb{R} \mid x \text{ is unlimited}\} = \emptyset, \tag{2.36}$$

$$^S\{x \in \mathbb{R} \mid x \approx 0\} = \{0\}, \tag{2.37}$$

$$^S\{x \in \mathbb{R} \mid x \not\approx 0\} = \mathbb{R} \setminus \{0\}. \tag{2.38}$$

> **!** The standardization axiom only says something about standard elements: The standard elements of $B := {}^S\{x \in A \mid \varphi(x)\}$ are precisely the standard elements of A that satisfy φ. This means that there can be non-standard elements of A that satisfy φ but are not in B (such as the nonzero infinitesimals in example (2.37)), as well as nonstandard elements of A that do not satisfy φ but are in B (such as the nonzero infinitesimals in example (2.38)).

The above examples are simple because one can immediately specify an equivalent internal description of the defined standard set. In general, however, this is not possible, or at least not so easily. We will get to know somewhat more sophisticated examples in the next section.

Before that, we note that the standardization schema implies the standard part axiom if we include the least-upper-bound property of $\mathbb{R}$ as an axiom in Section 2.5 (as is usually the case with an axiomatic introduction of real numbers). The standard part of a limited number $x_0 \in \mathbb{R}$ can then be defined by

$$\text{stp}(x_0) := \sup({}^S\{x \in \mathbb{R} \mid x \leq x_0\})$$

(see Section 3.5.4).

2.10.2 Implicit definitions via standardization

The external criteria from Section 2.8 are equivalent to the classical definitions, provided that their parameters are standard. By means of standardization, we can obtain *implicit* definitions from them, which are equivalent to the classical definitions without any restriction. Standardization ensures that those nonstandard elements are included that do not differ in any classic property from the externally characterized standard elements. We explain this in more detail for the concepts of convergence and continuity. The other concepts from Section 2.8 can be treated analogously using the respective external criteria.

In the following, $\mathbb{R}^{\mathbb{N}}$ denotes the set of all real-valued sequences and $\mathcal{F}$ the set of all real-valued functions f with $\text{dom}(f) \subseteq \mathbb{R}$. $\mathbb{R}^{\mathbb{N}}$ and $\mathcal{F}$ are standard sets, as are all Cartesian products of these sets with other standard sets, for example $\mathcal{F} \times \mathbb{R}$.

Convergence of sequences
We define the standard set

$$\mathcal{E}_1 := {}^S\{(f, a) \in \mathbb{R}^{\mathbb{N}} \times \mathbb{R} \mid \varphi_1(f, a)\},$$

where $\varphi_1(f, a)$ is the following external statement:

$$\text{For all } n \in \mathbb{N}, \text{ if } n \gg 1 \text{ then } f(n) \approx a.$$

Provided that f and a are standard, $\varphi_1(f, a)$ implies (by Theorem 14) that f converges to a in the classical sense. The general *nonstandard definition* for the convergence of sequences is now as follows:

Definition 7. A sequence $f \in \mathbb{R}^{\mathbb{N}}$ is called *convergent* if there is an $a \in \mathbb{R}$ such that $(f, a) \in \mathcal{E}_1$. For such an a, f is said to *converge to a*.

The convergence of sequences is thus defined by an internal statement with standard parameter $\mathcal{E}_1$. If a sequence is convergent, the number to which it converges is unique and is referred to as its *limit*. Uniqueness is guaranteed by the following theorem.

Theorem 40. *For any $f \in \mathbb{R}^{\mathbb{N}}$, there is at most one $a \in \mathbb{R}$ with $(f, a) \in \mathcal{E}_1$.*

Proof. Let f initially be standard. Any standard a with $(f, a) \in \mathcal{E}_1$ satisfies $\varphi_1(f, a)$ (according to the definition of $\mathcal{E}_1$). Therefore, given $n \gg 1$, for all standard a, a',

$$(f, a), (f, a') \in \mathcal{E}_1 \quad \Rightarrow \quad a \approx f(n) \approx a'.$$

Hence (since infinitely close standard numbers are equal), for all standard a, a',

$$(f, a), (f, a') \in \mathcal{E}_1 \quad \Rightarrow \quad a = a' \tag{2.39}$$

Again by transfer, (2.39) holds for *all* a, a'.
For general f, the assertion follows by transfer. □

Transfer was applied several times in this proof, always to internal statements (with standard parameters), i. e., legally.

Definition 7 is equivalent to the classical definition of convergence. For standard f, the equivalence follows from Theorem 14. The general case follows by transfer. We have thus found an equivalent internal description of $\mathcal{E}_1$, namely as the set of all $(f, a) \in \mathbb{R}^{\mathbb{N}} \times \mathbb{R}$ for which f converges to a according to the classical definition. However, Definition 7 does *not depend* on the possibility of describing $\mathcal{E}_1$ internally.

Continuity at a point

We define the standard set

$$\mathcal{E}_2 := {}^S\{(f, a) \in \mathcal{F} \times \mathbb{R} \mid \varphi_2(f, a)\},$$

where $\varphi_2(f, a)$ is the following external statement:

$$a \in \text{dom}(f) \text{ and for all } x \in \text{dom}(f), \text{ if } x \approx a \text{ then } f(x) \approx f(a).$$

The nonstandard definition of continuity is now as follows:

Definition 8. Let $f \in \mathcal{F}$ and $a \in \mathbb{R}$. Then f is *continuous* at a if $(f, a) \in \mathcal{E}_2$.

Definition 8 is equivalent to the conventional ε–δ definition of continuity. For standard f, the equivalence follows from Theorem 17. The general case follows by transfer. We have thus found an equivalent internal description of $\mathcal{E}_2$, namely as the set of all $(f, a) \in \mathcal{F} \times \mathbb{R}$ for which f is continuous at a according to the ε–δ definition. However, Definition 8 does *not* depend on the possibility of describing $\mathcal{E}_2$ internally.

The property described by $\varphi_2(f, a)$ is sometimes referred to as *S-continuity* of f at a. Continuity and S-continuity are therefore equivalent for standard parameters, but not in general.

Examples.
1. The standard function $x \mapsto \frac{1}{1+x^2}$ is continuous and S-continuous everywhere.
2. The standard function $x \mapsto x^2$ is continuous everywhere, but S-continuous only for limited x. For unlimited x, we have

$$x + \frac{1}{x} \approx x \quad \text{and} \quad \left(x + \frac{1}{x}\right)^2 = x^2 + 2 + \frac{1}{x^2} \neq x^2.$$

3. For $n \gg 1$, the nonstandard function $x \mapsto nx$ is continuous everywhere but nowhere S-continuous.

2.11 Outlook

2.11.1 Formalization and conservativeness

In this chapter, we have formulated the axioms of our theory colloquially, describing a universe that contains real numbers and sets (and possibly other objects). In order to state these axioms in a formal first-order language, we need a special symbol for each primitive notion of the theory. In addition to the already known symbols $0, 1, +, \cdot, <, \in$, we could, for example, introduce the unary relation symbols se ("is a set"), rn ("is a real number"), and st ("is standard").[16] Let $\mathcal{S}$ denote the set of all these nonlogical symbols, i.e.,

$$\mathcal{S} := \{0, 1, +, \cdot, \text{se}, \text{rn}, \text{st}, <, \in\}. \tag{2.40}$$

Instead of *statements*, one then speaks of *$\mathcal{S}$-formulas* or simply of *formulas* (if the reference to $\mathcal{S}$ is clear or unimportant). Formulas without free variables are also called *sentences*. How well-formed formulas can be constructed in a first-order language from logical and nonlogical symbols is defined recursively in the metalanguage (see, for example, [65], Section II.3).

---

[16] If ordered pairs are introduced axiomatically, another binary function symbol op must be added.

The pure set theories ZFC and IST get by with the symbol sets {∈} and {∈, st}, respectively. The ZFC axioms are therefore {∈}-sentences (short: ∈-sentences), the axioms from IST {∈, st}-sentences (short: ∈-st-sentences). For better readability, however, the axioms are usually given in an extended language with defined symbols such as ⊆, ∩, ∅. This is not a substantial extension of the language, as defined symbols can in principle be eliminated again.

For the sake of simplicity, we now assume that all symbols of $\mathcal{S}$, except for ∈ and st, are also *defined* by ∈-formulas, namely se(x) by $x = x$ (all objects should be sets) and rn(x) by $x \in \mathbb{R}$, where $\mathbb{R}$ is now a constant defined by one of the usual set-theoretic constructions of the real numbers in ZFC. The symbols 0, 1, +, ·, < are then to be defined accordingly. The set axioms from Sections 2.4.1 and 2.4.2 (except for the axiom of existence) then correspond exactly to the axioms of ZFC, and the axiom of existence, as well as the field and order axioms from Sections 2.5.1 and 2.5.2, is then provable in ZFC (where the axiom of choice is not needed).

If we add the nonstandard axioms from Section 2.6, but leave out the axiom of choice, the axiom system is equivalent to SPOT, which includes all axioms of ZF, as well as the eponymous axioms *Standard Part, nOntriviality*, and *Transfer* (see [104]). SP corresponds to our standard part axiom, O to our idealization axiom for real numbers,[17] and T to our transfer axiom.

SPOT is a *conservative* extension of ZF ([104], Theorem A). This means: Every ∈-sentence that is provable in SPOT is already provable in ZF. This result is interesting from a foundational point of view, as it shows that SPOT (and thus also the analysis presented here without the axiom of choice) is an epistemologically "harmless" extension of ZF (see Section 6.2.2). No assumptions are needed that imply conventional (i. e., internal) statements about sets that go beyond ZF (cf. on the other hand the remark on the ultrafilter theorem in Section 2.11.3). Conservativeness over ZF implies in particular consistency relative to ZF.

2.11.2 Countable idealization

In SPOT, the following principle of *countable idealization* is provable (cf. [104], Corollary 2.3):

Theorem 41 (countable idealization). *Let $\varphi(x,y)$ be an internal statement (optionally with parameters). Then for any countable standard set A the following statements are equivalent:*
1. *For every finite standard set $E \subseteq A$, there is an x such that $\varphi(x,y)$ holds for all $y \in E$.*
2. *There is an x such that $\varphi(x,y)$ holds for all standard $y \in A$.*

---

17 The axiom O is formulated in SPOT for natural numbers, but this makes no difference because of the Archimedean property of $\mathbb{R}$.

Since a finite subset of A is standard if and only if all its elements are standard,[18] this roughly means: A relation that is simultaneously satisfiable for finitely many arbitrary standard elements of A is simultaneously satisfiable for *all* standard elements of A—and vice versa. The prerequisite is that the relation is expressed by an internal statement.

In Nelson's internal set theory IST, the axiom schema of *idealization* is a generalization of Theorem 41 (see Section 3.5.3). The generalization consists of omitting the condition $E \subseteq A$ (with a countable standard set A).

2.11.3 Parameter-free standardization

In Section 2.10 we have seen how the concepts of convergence, continuity, derivative, integral, etc., can be implicitly defined by nonstandard criteria if the standardization axiom S of internal set theory is available (see Section 2.10.1). For the nonstandard definitions, even a weakened version of S is sufficient, where $\varphi(x)$ is not allowed to have parameters. Only such cases were used in Section 2.10. In [104] this "parameter-free" version of the standardization axiom is denoted by SN.[19] From a foundational point of view, it is significant that SPOT+SN (in contrast to SPOT+S) is still conservative over ZF (see [104], Theorem B). SPOT+S implies the Boolean prime ideal theorem and thus Tarski's ultrafilter theorem (every filter can be extended to an ultrafilter). Although the ultrafilter theorem is weaker than the axiom of choice, it cannot be proven in ZF.[20]

There are also theorems in conventional analysis whose proofs require the axiom of choice (at least its countable version), for example, the equivalence of ε–δ-continuity and continuity in terms of sequences. For such cases, a version of the axiom of choice must be added to the axioms. In [104] various extensions of SPOT are examined for this purpose.

---

[18] The proof of this is analogous to the proof of Theorem 1.1 in [160]. In the direction "⇐", it should be noted that for a countable set A the set of its finite subsets is also countable.

[19] SN is still equivalent to a more general version in which $\varphi(x)$ may have *standard* parameters (cf. [104], Lemma 6.1).

[20] For the role of the axiom of choice, see also Section 5.4.10.

3 An overview of nonstandard analysis

3.1 The history of nonstandard analysis

The history of analysis has been described many times, for example, in [42, 67, 205, 206]. The origins of calculus are generally traced back to antiquity, in some cases to the area and volume calculations of the Egyptians and Babylonians. Sonar therefore titled his book "3000 Years of Analysis".

In a narrower sense, analysis began with the work of Newton and Leibniz on calculus. They succeeded in linking the tangent and area problem (known today as the fundamental theorem of calculus), and their methods were applicable within a general framework.

The infinitely small played a decisive role in the first 200 years after Leibniz and Newton (but also before, for example, in Archimedes' heuristics or in Torricelli's reworking of Cavalieri's method of indivisibles; see [115]). If one calculates with such quantities as usual, the idea of the infinitely small implies the idea of the infinitely large, for example, in Leibniz's infinite bounded lines (see Section 5.2.2), which arise as reciprocals of infinitesimal quantities.

The counting numbers $1, 2, 3, \ldots$ must be continued up to infinite numbers in order to be able to calculate terms that are at infinite ranks in a sequence or a series, as Johann Bernoulli and Euler did, for example. The word *infinitesimal* (Neo-Latin: *infinitesimus*) is the ordinal form of the "number word" *infinitus*, so it actually means *infinitieth*.[1] The reference to the infinitely small is given by the fact that the infinitieth terms of a null sequence or a convergent series must be infinitely small, i.e., infinitesimal. However, the use of infinitesimals was also criticized from the outset (for example, by Nieuwentijt, Rolle, Berkeley).

Modern analysis, as taught at universities today, only became possible with the set-theoretical foundation of the real numbers by Cantor and Dedekind and Weierstrass' definition of limits, which made the use of infinitesimals superfluous. Another prerequisite was the development of modern quantifier logic, in particular by Frege, Peirce, and Peano, which made it possible to precisely define the meaning of a "limit".

The relief at this stroke of liberation can be seen in the following sentences with which Hilbert introduces his article *On the Infinite*:

> As a result of his penetrating critique, Weierstrass has provided a solid foundation for mathematical analysis. By elucidating many notions, in particular those of minimum, function, and differential quotient, he removed the defects which were still found in the infinitesimal calculus, rid it of all

---

[1] According to Probst, the term *pars infinitissima* still used by Mercator did not correspond to his actual use of infinitely small quantities, since the superlative *infinitissima* meant a *minimal* infinitely small quantity that cannot be further divided (similar to Cavalieri's indivisibles). Leibniz therefore switched to *infinitesima* and thus coined the term *infinitesimal* in 1673 (cf. [172], p. 103).

confused notions about the infinitesimal, and thereby completely resolved the difficulties which stem from that concept ([95], p. 161, as translated in [94]).

Immediately afterwards, however, Hilbert points out that although the infinitely large and the infinitely small have thus been successfully eliminated from analysis, "the infinite still appears in the infinite numerical series which defines the real numbers and in the concept of the real number system which is thought of as a completed totality existing all at once" (ibid.).

In fact, the banishment of infinitely small quantities from calculus came at a steep price, namely the systematic use of infinite sets. Hilbert mentioned infinite series and the completed totality of real numbers, i. e., infinite sets. According to the constructions of Cantor or Dedekind, every single real number is an infinite set (an infinite set of equivalent Cauchy sequences in $\mathbb{Q}$ or a Dedekind cut in $\mathbb{Q}$). Hardly anyone takes offense at this nowadays. Through years of practice in a standard math course, students are trained to perceive infinite sets as commonplace and infinitely small or infinitely large numbers as rather exotic. However, there does not appear to be an objective reason for this difference in perception (see Section 5.2).

Due to the unprecedented success of set theory in mathematics as a whole and in analysis with the newly invented real numbers and Weierstrass' concept of limits in particular, the infinitesimal quantities were soon no longer missed by most mathematicians. Cantor even referred to them as the "infinite cholera bacillus of mathematics" ([154], p. 505).

Despite this radical turn in analysis, infinitesimals as elements of non-Archimedean fields remained the subject of mathematical investigations, as Philip Ehrlich details in his article [69], for example, in Du Bois-Reymond, Veronese, Levi-Civita, Hahn, Artin and Schreier, Baer.

It is very easy to extend $\mathbb{R}$ or, more generally, any Archimedean field K to a non-Archimedean field K', for example, by adjunction of a new element Ω, which by definition is larger than all elements of K. Thus $K' := K(\Omega)$ is a non-Archimedean field that also contains infinitesimal numbers (for example, Ω^{-1}). Apart from the basic arithmetic operations and the ordering, however, nothing has been clarified for the new numbers. In the case $K = \mathbb{R}$, for example, it is not clear whether terms such as $\sqrt{\Omega}$, 2^{Ω}, $\sin(\Omega)$, $\Omega!$ or $\sum_{n=1}^{\Omega} n$ can be meaningfully defined. Field extensions of the type $\mathbb{R}(\Omega)$ are therefore uninteresting for analysis unless other definitions are added as, for example, in the generalized omega-adjunction in [130] (see Section 4.2.2).

Some progress can be made with more complex non-Archimedean field extensions. The Levi-Civita field is algebraically closed (if the imaginary unit is adjoint), thus allowing root extraction in particular [34].

In order to be useful for infinitesimal analysis, the field extensions must prove themselves in central tasks of analysis. Thus, the concept of integers should be meaningfully extended to the infinite numbers, which Bernoulli and Euler emphasized by speaking of "infinitieth" terms of a sequence (i. e., terms at an infinite rank). Applied to sequences of partial sums, this means that it should be possible to form sums with infinite bounds of summation. In particular, it should be possible to sum up an infinite number of infinitesimal numbers to a finite number for the definition of the integral.

The non-Archimedean fields listed above did not accomplish this. Fraenkel summarized the disappointing situation in 1928 as follows:

> The types of infinitely small quantities considered so far, some of which have been carefully justified, have proven to be completely useless for solving even the simplest and most basic problems of infinitesimal calculus (such as proving the mean value theorem of differential calculus or defining the definite integral) ([76], p. 116, own translation).[2]

Schmieden and Laugwitz achieved a partial success in 1958 with their omega calculus, which provides infinitely small and infinitely large numbers and allows the treatment of all essential parts of classical analysis as well as novel *nonstandard functions* (such as delta functions) [190]. However, the number system of omega numbers contains nontrivial zero divisors and is only partially ordered. The omega numbers can therefore not be reconciled with the idea of a linear continuum.[3]

The breakthrough came in 1961 with Abraham Robinson's *Non-standard Analysis* [181] building upon work by Hewitt [92] and Łoś [143]. Nonstandard models of analysis, as used by Robinson, are not unique. Their existence results from the compactness theorem which was Robinsons's original approach (see Section 3.3.3) or from a construction using ultrafilters (see Section 3.3.1). Today, the domain of such nonstandard models is usually denoted by $^*\mathbb{R}$, and its elements are called *hyperreal numbers*. The term "hyperreal" goes back to Hewitt ([92], p. 74).

The sets $^*\mathbb{R}$ and $\mathbb{R}$ are elementarily equivalent with respect to a first-order language that contains a symbol for each real number and for each relation on $\mathbb{R}$ (see Section 3.3.3). Thus $^*\mathbb{R}$ is not only a non-Archimedean extension field of $\mathbb{R}$, but satisfies a much more far-reaching transfer principle, which can be seen as a formalization of Leibniz's law of continuity, as Robinson noted in his 1966 book (see [182], p. 266).[4] Further work by Robinson, Zakon, and Luxemburg generalized the concept to so-called superstructures of arbitrary infinite sets and made it applicable to many other areas of mathematics [182, 186]. The construction of $^*\mathbb{R}$ or more generally of nonstandard extensions of superstructures uses nonprincipal ultrafilters whose existence can be proved (nonconstructively) with Zorn's lemma.

---

2 In the original German: "Die bisher in Betracht gezogenen und teilweise sorgfältig begründeten Arten unendlich kleiner Grössen haben sich zur Bewältigung auch nur der einfachsten und grundlegendsten Probleme der Infinitesimalrechnung (etwa zum Beweis des Mittelwertsatzes der Differentialrechnung oder zur Definition des bestimmten Integrals) als völlig unbrauchbar erwiesen."
3 The same number system was already considered by Peano in 1910 (see footnote 7).
4 For a comparison of Leibniz and nonstandard analysis, see also [113].

In the 1970s, two axiomatic approaches to nonstandard analysis were proposed independently of each other: the *internal set theory* by Edward Nelson [160] and the set theory by Karel Hrbaček [100]. Both proposals extend the language and the axiom system of classical set theory so that nonstandard sets appear in the resulting set universe in addition to the standard sets. However, both extensions are conservative over ZFC (see Sections 3.5 and 3.6).[5] As the extensions are based directly on the foundation of mathematics, i. e., set theory, the nonstandard concepts are very generally applicable from the outset. Nelson's internal set theory has had the greatest influence on practiced mathematics to date.

In addition to the model-theoretical and axiomatic approaches to nonstandard analysis, from the 1980s onwards there was an advance from category theory that goes back to Francis William Lawvere [131, 132]: *smooth infinitesimal analysis*. Infinitesimals are here nil-square elements, i. e., elements whose square is exactly 0 without having to be 0 themselves. In order for such elements to be possible, intuitionistic logic is used, in which (in contrast to classical logic) the *tertium non datur* does not apply. Therefore $\neg \varepsilon \neq 0$ does not imply $\varepsilon = 0$.

In this chapter, we take a closer look at three approaches:
- the omega calculus according to Schmieden and Laugwitz, as it is still purely constructive and comprehensible with elementary means and is a good preparation for the Robinson approach,
- Robinson's model-theoretical approach, as it is still the most commonly practiced by users of nonstandard analysis,
- Nelson's internal set theory as an example of an axiomatic approach, since it is very popular with users due to its simple axiomatics and since it is particularly suitable as a starting point for philosophical and foundational discussions.

Other axiomatic approaches (external and relative set theories) are outlined in less detail. For an introduction to smooth infinitesimal analysis, we refer to [29].

3.2 Omega calculus according to Schmieden and Laugwitz

Cantor defined the real numbers as equivalence classes of Cauchy sequences in $\mathbb{Q}$. Two such sequences are equivalent (i. e., they represent the same real number) if their difference is a null sequence. Similarly, Schmieden and Laugwitz defined their omega numbers as equivalence classes of sequences in an ordered field K (for example, $K = \mathbb{Q}$ or $K = \mathbb{R}$).[6] The difference to Cantor's definition is, firstly, that not only Cauchy sequences, but all sequences in K are allowed, and secondly, that the equivalence relation is stricter.

---

[5] In addition to these original proposals, there are other variants. A comprehensive overview of the axiomatic approaches to nonstandard analysis can be found in the monograph [110].

[6] See [129]. Originally, Schmieden and Laugwitz had defined their calculus for Ω-rational numbers [190].

Two sequences only represent the same omega number if they agree *everywhere* [190] or at least *almost everywhere* [129]. "Almost everywhere" here means for all indices, possibly with finitely many exceptions.[7]

For example, the sequences $(\frac{1}{n})$ and $(\frac{1}{2n})$ in Cantor's construction both represent the real number zero. But they represent two different omega numbers, both of which are nonzero and of which the first is twice as large as the second.

The description of the omega calculus in this section is essentially based on [129]. In the following, we suppose that K is an Archimedean field.

3.2.1 Definition of omega numbers

A subset of $\mathbb{N}$ is called *cofinite* if its complement in $\mathbb{N}$ is finite. In order to be able to conveniently formulate properties that hold almost everywhere, we introduce the system Cof of all cofinite subsets of $\mathbb{N}$, i.e., we define

$$\mathrm{Cof} := \{E \in \mathcal{P}(\mathbb{N}) \mid \mathbb{N} \setminus E \text{ is finite}\}.$$

In the following, we will often use that intersections and supersets of cofinite sets are cofinite again, i.e., that for all sets A, B,

$$A, B \in \mathrm{Cof} \quad \Rightarrow \quad A \cap B \in \mathrm{Cof}, \tag{3.1}$$

$$A \in \mathrm{Cof}, A \subseteq B \subseteq \mathbb{N} \quad \Rightarrow \quad B \in \mathrm{Cof}. \tag{3.2}$$

These properties of Cof follow directly from the definition.[8]

Let $K^\mathbb{N}$ denote the set of sequences in K and let the binary relation $\sim_{\mathrm{Cof}}$ on $K^\mathbb{N}$ be defined by

$$a \sim_{\mathrm{Cof}} b \quad :\Leftrightarrow \quad \{n \in \mathbb{N} \mid a_n = b_n\} \in \mathrm{Cof} \tag{3.3}$$

for all $a, b \in K^\mathbb{N}$ with $a = (a_n)$ and $b = (b_n)$.

It is easy to check that $\sim_{\mathrm{Cof}}$ is an equivalence relation on $K^\mathbb{N}$. Reflexivity and symmetry are clear. For transitivity, one needs (3.1) and (3.2).

This allows us to define the set $^\Omega K$ of *omega numbers* (based on K) as $K^\mathbb{N}/\sim_{\mathrm{Cof}}$. That is, we define

$$^\Omega K := \{[a] \mid a \in K^\mathbb{N}\},$$

---

[7] Before Schmieden and Laugwitz, in 1910 Peano had already investigated the set of all eventually equal sequences over $\mathbb{R}$ as a number system containing infinitesimals (see [169], reprinted in [168]).

[8] Together with the property that Cof is nonempty and does not contain the empty set, they characterize Cof as a *filter* on $\mathbb{N}$ (see Definition 11).

where $[a] := \{b \in K^\mathbb{N} \mid b \sim_{\text{Cof}} a\}$ denotes the equivalence class of $a \in K^\mathbb{N}$ with respect to $\sim_{\text{Cof}}$. For $a = (a_n) \in K^\mathbb{N}$, we say that (a_n) *represents* the omega number $[a]$. Laugwitz designates the omega number represented by (a_n) as a_Ω or α (and analogously for other Latin letters and their Greek equivalents).

The mapping $\rho: K \to {}^\Omega K$, which assigns to each constant sequence in K its equivalence class with respect to $\sim_{\text{Cof}}$, is injective because two different constant sequences do not agree anywhere (and thus not almost everywhere). The field K is *embedded* in ${}^\Omega K$ by ρ and can thus be understood as a subset of ${}^\Omega K$ (by identifying $a \in K$ with $\rho(a)$). Therefore we can continue to use designations for elements from K in ${}^\Omega K$ and, for example, simply write 1 again instead of $[(1,1,1,\dots)]$. This practice is common for extensions of number systems, for example, $\mathbb{N} \subset \mathbb{Z} \subset \mathbb{Q} \subset \mathbb{R}$, where 1 is used to denote the natural, whole, rational, and real number 1.

The mapping ρ is not surjective because the sequence $(1,2,3,\dots)$ does not agree almost everywhere with any constant sequence. Therefore, the number

$$\Omega := [(1,2,3,\dots)]$$

is not in the range of ρ. The step from K to ${}^\Omega K$ is hence a proper extension.

3.2.2 Extension of relations and functions

For every m-ary relation R on K, there is a corresponding m-ary relation ${}^*R$ on ${}^\Omega K$ defined by

$$(\alpha_1, \dots, \alpha_m) \in {}^*R \quad :\Leftrightarrow \quad \{n \in \mathbb{N} \mid (a_{1,n}, \dots, a_{m,n}) \in R\} \in \text{Cof}. \qquad (3.4)$$

Here, $(a_{j,n})_{n \in \mathbb{N}}$ represents α_j (for $j = 1, \dots, m$). Due to (3.1) and (3.2), ${}^*R$ is well defined, i.e., does not depend on the choice of the representing sequences.

As usual, the concept of function is traced back to the concept of relation. Using (3.1), (3.2), and (3.4), it can be shown that if f is an m-ary function from $D \subseteq K^m$ to K, then ${}^*f$ is an m-ary function from ${}^*D$ to ${}^\Omega K$ and

$${}^*f(\alpha_1, \dots, \alpha_m) = \beta \quad :\Leftrightarrow \quad \{n \in \mathbb{N} \mid f(a_{1,n}, \dots, a_{m,n}) = b_n\} \in \text{Cof}. \qquad (3.5)$$

The embedding function ρ is an injective homomorphism for every m-ary relation R on K, i.e., for all $a_1, \dots, a_m \in K$,

$$(a_1, \dots, a_m) \in R \quad \Rightarrow \quad (\rho(a_1), \dots, \rho(a_m)) \in {}^*R.$$

If K is understood as a subset of ${}^\Omega K$ (by embedding via ρ), then ${}^*R$ is an extension of R (and correspondingly ${}^*f$ is an extension of the function f).

If no misunderstandings are to be expected, the asterisk is therefore omitted from the function and relation symbols, especially for $+, \cdot, <$. In this way we have

$$\alpha < \beta \quad :\Leftrightarrow \quad \{n \in \mathbb{N} \mid a_n < b_n\} \in \text{Cof}, \tag{3.6}$$

$$\alpha + \beta = \gamma \quad :\Leftrightarrow \quad \{n \in \mathbb{N} \mid a_n + b_n = c_n\} \in \text{Cof}, \tag{3.7}$$

$$\alpha \cdot \beta = \gamma \quad :\Leftrightarrow \quad \{n \in \mathbb{N} \mid a_n \cdot b_n = c_n\} \in \text{Cof}. \tag{3.8}$$

For unary relations (i.e., sets) such as $^*\mathbb{N}$ and $^*\mathbb{Q}$, the asterisk is still required to distinguish the extended sets from the (embedded) original sets $\mathbb{N}$ and $\mathbb{Q}$.

Since the calculation rules for addition and multiplication are transferred from the terms of the representing sequences to the omega numbers, we find that $^\Omega K$ is a commutative ring with identity.

However, $^\Omega K$ contains nontrivial zero divisors. For example,

$$[(1,0,1,0,\dots)] \cdot [(0,1,0,1,\dots)] = 0,$$

without one of the factors being zero. Hence, $^\Omega K$ is not a field and cannot be extended to a field. The reciprocal α^{-1} only exists for numbers α for which the representing sequences are nonzero almost everywhere.

Furthermore, $^\Omega K$ is only partially ordered. The relation $<$ on $^\Omega K$ is irreflexive and transitive, but not total. For example, the omega number $(-1)^\Omega = [(-1,1,-1,1,\dots)]$ is neither less than, nor equal to, nor greater than zero, since none of these conditions hold almost everywhere for the representing sequence. Despite these algebraic deficits compared to K, a relatively far-reaching infinitesimal calculus is possible with omega numbers, as demonstrated in [129].

3.2.3 Infinite and infinitesimal omega numbers

Definition 9. An omega number ξ is said to be
- *positive infinite* ($\xi \gg 1$) if $\xi > a$ for all $a \in K$,
- *negative infinite* if $-\xi$ is positive infinite,
- *infinite* if it is positive or negative infinite,
- *finite* if there is an $a \in K$ with $|\xi| \leq a$,[9]
- *infinitesimal* or *infinitely small* ($\xi \approx 0$) if $|\xi| < a$ for all positive $a \in K$.[10]

---

[9] In the systems of omega or hyperreal numbers, it is common to use the historical terms *finite* and *infinite*. However, some authors prefer the terms *limited* and *unlimited* for the hyperreals as well (see, e.g., [85]).
[10] For Laugwitz, *infinitesimal* implies nonzero. For the sake of consistent terminology, we include zero in the infinitesimal numbers here.

Two omega numbers ξ, η are said to be *infinitely close* ($\xi \approx \eta$) if their difference is infinitesimal.

From the definition, it follows that Ω is positively infinite and Ω^{-1} is infinitesimal.

Since sequences in K are functions from $\mathbb{N}$ to K, they can be extended to functions from $^*\mathbb{N}$ to $^\Omega K$. From a sequence $(a_n)_{n \in \mathbb{N}}$, we get the extension $(a_n)_{n \in {^*\mathbb{N}}}$, where $a_\nu = [(a_{n_j})_{j \in \mathbb{N}}]$ for each $\nu \in {^*\mathbb{N}}$ represented by $(n_j)_{j \in \mathbb{N}}$. Specifically for $\nu = \Omega$, we have $a_\Omega = [(a_j)_{j \in \mathbb{N}}]$ as a justification for Laugwitz' naming convention for omega numbers.

For historical reasons and following Leibniz's notation for differentials, infinitesimal omega numbers are also designated as dx, dy, etc.

3.2.4 Hyperfinite sequences and sums

The nonstandard definition of the integral requires a generalization of the terms *finite sequence* and *finite sum*. An infinite number of omega numbers should be allowed as terms in a sequence or a sum, but in such a way that essential properties of finite sequences or sums are preserved. This leads us to the concept of *hyperfinite* sequences and sums.

A finite sequence $(x_k)_{1 \leq k \leq n}$ in K is a mapping

$$\{k \in \mathbb{N} \mid 1 \leq k \leq n\} \to K, \quad k \mapsto x_k.$$

This is usually written as $(x_1, \ldots, x_n)$. The natural number n is called the *length* of the sequence.

For the generalization, let $\nu, \kappa \in {^*\mathbb{N}}$ be represented by (n_j) and (k_j), respectively (which are sequences in $\mathbb{N}$).

A *hyperfinite sequence* $(\xi_\kappa)_{1 \leq \kappa \leq \nu}$ in $^\Omega K$ is a mapping

$$\{\kappa \in {^*\mathbb{N}} \mid 1 \leq \kappa \leq \nu\} \to {^\Omega K}, \quad \kappa \mapsto \xi_\kappa := [(x_{j,k_j})],$$

where $(x_{j,k})_{1 \leq k \leq n_j}$ is a finite sequence in K for each $j \in \mathbb{N}$. The omega number ν is called the *length* of the hyperfinite sequence. A hyperfinite sequence of length ν is thus defined by an infinite sequence of finite sequences of length n_j. The range of a hyperfinite sequence is called a *hyperfinite set*.

Hyperfinite sequences are a special case of *internal mappings* (see Section 3.4). The definition given above results from the fact that internal objects over $^\Omega K$ (e. g., sets, functions, relations) are defined by sequences of corresponding objects over K (see [129], pp. 79–80). As a result, essential properties are transferred from the objects over K to the corresponding internal objects over $^\Omega K$.

For example, every hyperfinite sequence in $^\Omega K$ (as well as every finite sequence in K) has a maximal term. To prove this, choose for each j in the above definition the k_j

for which x_{j,k_j} is maximal in $(x_{j,1}, \ldots, x_{j,n_j})$. Then the corresponding ξ_κ is maximal in the hyperfinite sequence $(\xi_\kappa)_{1\leq\kappa\leq\nu}$.

Just as we can sum up the terms of a finite sequence in K, we can do the same with the terms of a hyperfinite sequence of omega numbers. The sum of the terms of a hyperfinite sequence $(\xi_\kappa)_{1\leq\kappa\leq\nu}$ is defined by

$$\sum_{k=1}^{\nu} \xi_k := \left[\left(\sum_{k=1}^{n_j} x_{j,k}\right)\right].$$

This is called a *hyperfinite sum*. More generally, for $\mu \leq \nu$ one defines

$$\sum_{k=\mu}^{\nu} \xi_k := \left[\left(\sum_{k=m_j}^{n_j} x_{j,k}\right)\right].$$

Similarly, one can define *hyperfinite products*.

3.2.5 Central concepts of elementary analysis with omega numbers

We now assume the main case of interest $K = \mathbb{R}$.

Since two different real numbers cannot be infinitely close, there is at most one infinitely close real number for each omega number ξ. If such a number exists, it is called the *standard part* of ξ and is denoted by $\mathrm{stp}(\xi)$. A necessary (but not sufficient) condition for the existence of the standard part is that ξ is finite. The standard part of ξ exists if and only if the representing sequence (x_n) converges in the conventional sense. In this case,

$$\mathrm{stp}(\xi) = \lim_{n\to\infty} x_n.$$

The terms continuity, derivative, integral, and limit for real functions can be defined with omega numbers (and analogously with hyperreal numbers) as follows (cf. [129], pp. 25–39 and 54–68, for the integral also pp. 139–140):

Definition 10. Consider $f : D \to \mathbb{R}$ with $D \subseteq \mathbb{R}$. Then
1. f is said to be *continuous* at $x \in D$, provided for all $dx \approx 0$ with $x + dx \in {}^*D$,

$$f(x + dx) \approx f(x).$$

2. f is said to be *differentiable* at $x \in D$ with *derivative* $f'(x) \in \mathbb{R}$, provided ${}^*D$ contains points infinitely close but not equal to x and for all $dx \approx 0$ that are not zero divisors[11] and for which $x + dx \in {}^*D$,

---

[11] In the case of hyperreal numbers, this means $dx \neq 0$.

$$\frac{f(x+dx)-f(x)}{dx} \approx f'(x).$$

3. Let $D = [a,b]$ be a nondegenerate interval, $(\xi_k)_{0 \le k \le \nu}$ an infinitely fine partition of D (i. e., a strictly monotonically increasing hyperinfinite sequence with $\xi_0 = a$, $\xi_\nu = b$, and $\xi_k - \xi_{k-1} \approx 0$) and $(\hat{\xi}_k)_{1 \le k \le \nu}$ a choice of intermediate points (i. e., a hyperfinite sequence with $\xi_{k-1} \le \hat{\xi}_k \le \xi_k$). If the standard part of the Riemann sum

$$\sum_{k=1}^{\nu} f(\hat{\xi}_k)(\xi_k - \xi_{k-1})$$

exists and is the same for all infinitely fine partitions and any choice of intermediate points, then it is called the *definite integral* $\int_a^b f(x)\,dx$ and f is said to be *integrable* from a to b.

4. Let $D = \mathbb{N}$. The sequence f is said to *converge* to the *limit* $a \in \mathbb{R}$, provided for all infinitely large $\nu \in {}^*\mathbb{N}$,

$$f(\nu) \approx a.$$

3.2.6 Limitations

A great strength of Robinson's nonstandard analysis is its far-reaching *transfer principle*, which states that every first-order statement that is valid in the real numbers is also valid in the hyperreal numbers if all functions and relations are replaced by their *-extensions (see Section 3.3.2).

Given the differences found between K and ${}^\Omega K$ (existence of nontrivial zero divisors, no total order), it is clear that a transfer principle for omega numbers cannot apply to all first-order formulas. In fact, only those formulas are generally transferable that contain only $\wedge$, $\forall$, $\exists$ as logical symbols.[12] The transfer principle for omega numbers is therefore only of limited benefit in practice.

In his omega calculus in [129], Laugwitz dispenses with the explicit formulation of a transfer principle and instead refers directly to the construction of omega numbers in proofs. An introduction of formal languages is therefore not necessary at this stage.

The reason for the restricted transfer principle for omega numbers is that a subset of $\mathbb{N}$ can be infinite and its complement in $\mathbb{N}$ can also be infinite. Neither set therefore belongs to Cof. In relation to the terms of a sequence (a_n), which represents the omega number α, this means: If a statement does not hold for almost all a_n, it does not follow that the negation of the statement holds for almost all a_n. Correspondingly, if a disjunction of two statements is true for almost all a_n, it does not follow that one of the two statements is true for almost all a_n. In the concrete example $(-1)^\Omega$, represented

---

[12] A more detailed investigation can be found in [124], pp. 31–34.

by the sequence $((-1)^n)$: For almost all n (in this case even for all n), the disjunction $a_n = -1 \lor a_n = 1$ is true, but it is neither $a_n = -1$ for almost all n, nor $a_n = 1$ for almost all n. Therefore, $(-1)^\Omega$ is neither equal to -1 nor equal to 1.

Furthermore, $(-1)^\Omega$ is also an example of a finite omega number that has no standard part because the sequence $((-1)^n)$ is not convergent. Since not every finite omega number has a standard part, additional considerations must sometimes be made in proofs, for example, by exploiting the fact that bounded sequences have convergent subsequences (cf. [129], pp. 56–58).

These disadvantages of omega numbers can be remedied by defining the equivalence relation on $K^\mathbb{N}$ not with the filter Cof, but with a larger filter, a so-called *nonprincipal* or *free ultrafilter* on $\mathbb{N}$ (see Section 3.3.1). Such an ultrafilter $\mathcal{U}$ contains for each subset of $\mathbb{N}$ either the set itself or its complement, making $^\Omega K$ an ordered field. Then $(-1)^\Omega$ is, for example, either equal to 1 or equal to -1 because (depending on $\mathcal{U}$) either the set of all even indices or the set of all odd indices belongs to $\mathcal{U}$. Laugwitz considers such field extensions, for example, in [130].

3.3 The hyperreal numbers

The hyperreal numbers form an ordered, non-Archimedean extension field of $\mathbb{R}$. Such a field is obtained by a set-theoretic construction using free ultrafilters (as indicated in the previous section) or as a consequence of the compactness theorem of first-order logic. We briefly introduce both ways.

3.3.1 Field extension via ultrafilter

When proving statements about omega numbers, the properties (3.1) and (3.2) of Cof were repeatedly employed, i.e., that with two sets their intersection also belongs to Cof and that with a set A the subsets of $\mathbb{N}$ that are supersets of A also belong to Cof. In addition, Cof is not empty and does not contain the empty set. These properties make Cof a so-called *filter* on $\mathbb{N}$. It is also called the *Fréchet filter on* $\mathbb{N}$.

More generally, filters are defined on an arbitrary nonempty set J.[13] In the following, let J be any nonempty set.

Definition 11. 1. A nonempty family $\mathcal{F}$ of subsets of J is called a *filter on J* if it has the following properties:
   (a) $\emptyset \notin \mathcal{F}$.
   (b) For all A, B, if $A, B \in \mathcal{F}$, then $A \cap B \in \mathcal{F}$.

---

[13] Index sets other than $\mathbb{N}$ are used, for example, to prove the existence of nonstandard embeddings with certain additional properties (see Sections 3.4.10 and 3.4.11).

(c) For all A, B, if $A \in \mathcal{F}$ and $A \subseteq B \subseteq J$, then $B \in \mathcal{F}$.
2. A filter $\mathcal{F}$ is called an *ultrafilter* if for every $X \subseteq J$,

$$X \in \mathcal{F} \quad \text{or} \quad J \setminus X \in \mathcal{F}. \tag{3.9}$$

3. A filter $\mathcal{F}$ is said to be *free* or *nonprincipal* if

$$\bigcap \mathcal{F} = \emptyset. \tag{3.10}$$

Since $\mathcal{F}$ is not empty, property 1c in the Definition 11 implies $J \in \mathcal{F}$. Since the intersection of all cofinite subsets of $\mathbb{N}$ is empty, Cof (and thus every filter comprising Cof) is free. However, Cof is not an ultrafilter, as it does not satisfy (3.9) for all $X \subseteq \mathbb{N}$.

The next theorem shows that ultrafilters are the largest possible filters (see [106], p. 74):

Theorem 42. *Let $\mathcal{F}$ be a filter on J. Then the following statements are equivalent:*
1. *$\mathcal{F}$ is an ultrafilter.*
2. *$\mathcal{F}$ is maximal, i. e., there is no filter $\mathcal{G}$ on J for which $\mathcal{F} \subset \mathcal{G}$.*

The following theorem can then be proved using Zorn's lemma (see [106], p. 75).

Theorem 43 (Tarski's ultrafilter theorem). *Every filter can be extended to an ultrafilter.*

This theorem is sometimes also called *ultrafilter lemma*.

3.3.2 Transfer principle for hyperreal numbers

According to Theorem 43, Cof can be extended to an ultrafilter $\mathcal{U}$, which is then a free ultrafilter on $\mathbb{N}$. If we now modify all definitions from Section 3.2 by replacing Cof with $\mathcal{U}$, we can prove a transfer principle that applies to all first-order statements.[14] The phrase "almost everywhere" now means "for all n of a set that belongs to $\mathcal{U}$". Following Robinson, we denote the extension field $\mathbb{R}^{\mathbb{N}}/\sim_{\mathcal{U}}$ by $^*\mathbb{R}$. The definition of $^*\mathbb{R}$ therefore depends on the ultrafilter $\mathcal{U}$. We will come back to this fact in Section 6.2.1. The elements of $^*\mathbb{R}$ are called *hyperreal numbers*.

We need a suitable formal first-order language $\mathcal{L}^S$ (where S the set of its nonlogical symbols) in which the transferable sentences can be formulated. Let S be the symbol set that contains a constant for each real number and a relation symbol for each relation on $\mathbb{R}$. Such a symbol set cannot be specified effectively. For the sake of simplicity, suppose that the real numbers and the relations on $\mathbb{R}$ are themselves the symbols in S. This means that in concrete examples, all familiar symbols for real numbers or relations on

---

14 For the designations and concepts from first-order logic used in the following, see, for example, [65].

ℝ from background set theory can be used in S-formulas, for example, 10, $-\frac{2}{3}$ or π as constants, $\mathbb{N}, \mathbb{Z}, \mathbb{Q}$ as unary relation symbols, $<, >, \leq, \geq$ as binary relation symbols, $+, \cdot$ as ternary relation symbols. In case of doubt, it must then be stated whether a character string is to be read as a formal S-formula or as a statement of background set theory.

We denote S-structures with $\mathfrak{A}, \mathfrak{B}$, etc., and mark the interpretations of the symbols from S in these structures with the corresponding superscript. For example, $a^{\mathfrak{A}}$ denotes the interpretation of the constant a in $\mathfrak{A}$. Let $\mathfrak{A}$ be the structure with domain $\mathbb{R}$ and $a^{\mathfrak{A}} := a$ for all constants a and $R^{\mathfrak{A}} := R$ for all relation symbols R. Further, let $\mathfrak{B}$ be the structure with domain $^*\mathbb{R}$ and $a^{\mathfrak{B}} := \rho(a) = [(a)_{n \in \mathbb{N}}]$ for all constants a and $R^{\mathfrak{B}} := {}^*R$ for all relation symbols R.

The following theorem provides the connection between statements about hyperreal numbers and the corresponding statements about the sequences representing them (cf. [66], pp. 100–102).

Theorem 44. *Let φ be an S-formula with at most the free variables $x_1, \ldots, x_m$. Then for all $\alpha_1, \ldots, \alpha_m \in {}^*\mathbb{R}$ with $\alpha_j = [(a_{j,n})_{n \in \mathbb{N}}]$ for $j = 1, \ldots, m$,*

$$\{n \in \mathbb{N} \mid \mathfrak{A} \models \varphi[a_{1,n}, \ldots, a_{m,n}]\} \in \mathcal{U} \quad \Leftrightarrow \quad \mathfrak{B} \models \varphi[\alpha_1, \ldots, \alpha_m].$$

Here $\models$ means "models". In simple terms, Theorem 44 states: A statement is true for hyperreal numbers if and only if it is true almost everywhere in relation to the terms of the representing sequences.

The following transfer principle can be derived from Theorem 44:

Theorem 45 (transfer principle for hyperreal numbers). *For every S-sentence φ,*

$$\mathfrak{A} \models \varphi \quad \Leftrightarrow \quad \mathfrak{B} \models \varphi.$$

If $\varphi^{\mathfrak{A}}$ and $\varphi^{\mathfrak{B}}$ denote the sentence φ interpreted in $\mathfrak{A}$ and $\mathfrak{B}$, respectively, Theorem 45 can be written even shorter as $\varphi^{\mathfrak{A}} \Leftrightarrow \varphi^{\mathfrak{B}}$.

Since the constants and relation symbols from S are interpreted trivially in $\mathfrak{A}$ ($a^{\mathfrak{A}} = a$ and $R^{\mathfrak{A}} = R$), the formal sentence φ and its interpretation $\varphi^{\mathfrak{A}}$ differ visually only in that the more informal language of background set theory is used for the latter (and the signs $\Rightarrow$ and $\Leftrightarrow$ instead of $\rightarrow$ and $\leftrightarrow$). The difference is further leveled by the fact that an informal notation is usually also permitted to specify concrete formal sentences, for example, by writing $x + y = z$ instead of $+xyz$ or $x < y$ instead of $< xy$.

3.3.3 Field extension via compactness theorem

Robinson's original approach to nonstandard analysis in [181] uses the compactness theorem. This theorem from first-order logic states that a set Φ of formulas is satisfiable (i.e., has a model) if and only if every finite subset of Φ is satisfiable (see [65], p. 84). In order to derive the existence of a suitable structure for nonstandard analysis, a formal first-order language with an uncountable number of constants and relation symbols is employed.[15]

As in Section 3.3.2, the symbol set S contains a constant for each real number $a \in \mathbb{R}$ and a relation symbol for each relation R on $\mathbb{R}$. For the sake of simplicity, suppose again that the real numbers and the relations themselves are the symbols in S.

Let $\mathfrak{A}$ be the structure with domain $\mathbb{R}$ and the trivial interpretation of the constants and relation symbols ($a^{\mathfrak{A}} = a$ or $R^{\mathfrak{A}} = R$) and $\text{Th}(\mathfrak{A})$ the set of all S-sentences that hold in $\mathfrak{A}$.

Further, let Φ be the set of formulas $\text{Th}(\mathfrak{A}) \cup \{x > 1, x > 2, x > 3, \ldots\}$. Every finite $\Phi_0 \subseteq \Phi$ is satisfiable, for example, by an interpretation $(\mathfrak{A}, \beta_1)$ with sufficiently large $\beta_1(x)$.[16] According to the compactness theorem, there is then also a model $(\mathfrak{B}, \beta)$ of Φ. Designate its domain with $^*\mathbb{R}$. Because of $\mathfrak{B} \models \text{Th}(\mathfrak{A})$, $\mathfrak{A}$ and $\mathfrak{B}$ are elementary equivalent. With $\Omega := \beta(x)$, however, $^*\mathbb{R}$ also contains an infinitely large element.

The set $\mathbb{R}$ can be embedded in $^*\mathbb{R}$ (via $a \mapsto a^{\mathfrak{B}}$) and we have

$$R(a_1, \ldots, a_n) \Leftrightarrow R^{\mathfrak{B}}(a_1^{\mathfrak{B}}, \ldots, a_n^{\mathfrak{B}})$$

for all $a_1, \ldots, a_n \in \mathbb{R}$ and for all n-ary relations R on $\mathbb{R}$ (because of $\mathfrak{B} \models \text{Th}(\mathfrak{A})$). Hence, $\mathfrak{A}$ can be understood as a substructure of $\mathfrak{B}$ and $\mathfrak{B}$ as an elementary extension of $\mathfrak{A}$.

This gives us the structure of the hyperreal numbers without having to construct it explicitly. However, the procedure in Section 3.3.1 did not really construct the extension structure either, as the ultrafilter used could not be specified explicitly.

3.4 Nonstandard analysis in superstructures

The means from Section 3.3 are sufficient for analysis with standard functions and relations on $\mathbb{R}$, to a certain extent even for the treatment of nonstandard functions (see [181], p. 437 f.). If nonstandard methods are to be used in a more general context (for example, in topology, functional analysis or stochastics), the means must be extended. In particular, the terms *internal set* and *hyperfinite set* must be generalized.

So-called *superstructures* are considered as the domains of discourse. These are built analogously to the von Neumann hierarchy (see Section 5.4.2) up to V_ω, but in a set theory with *urelements*, i. e., objects that are not sets. The lowest level of the hierarchy is therefore not the empty set, but an infinite set S of urelements.[17] The elements of S are also called *atoms* in the following.

---

[15] The omission of function symbols in S does not imply any significant restriction (see [65], Section VIII.1).

[16] Here β_1 is a so-called *assignment*, i. e., a mapping from the set of variables into the domain of the structure.

[17] The assumption of an infinite set of urelements is consistent relative to ZFC (see [106], p. 250).

In order to build a nonstandard theory for superstructures, a formal language is introduced whose symbol set (in addition to the relation symbol ∈) contains a constant for each element of the superstructure. Since functions and relations on S (as well as on the higher hierarchy levels) are already contained in the superstructure (and thus represented as constants in the symbol set), function symbols and further relation symbols can be dispensed with.

The embedding of the structure of the real numbers in the more comprehensive structure of the hyperreal numbers in Section 3.3.3 now corresponds to the embedding of the superstructure of S in a more comprehensive superstructure. This leads to the concept of *nonstandard embedding*. The transfer principle then no longer applies to all first-order sentences, but at least to those in which the quantifications are restricted by constants or variables (*transitively bounded sentences*).

The following presentation is based on [224].

3.4.1 Transitively bounded formulas

Let S be a symbol set that contains the binary relation symbol ∈ and otherwise only constants. A formula is called *transitively bounded* if it is constructed according to the following rules:

1. Every quantifier-free formula is transitively bounded.
2. If φ is a transitively bounded formula, x a variable and t a constant or a variable different from x, then the formulas $(\forall x\, (x \in t \rightarrow \varphi))$ and $(\exists x\, (x \in t \wedge \varphi))$ are transitively bounded. In these cases we also write

$$\forall x \in t\, \varphi \quad \text{and} \quad \exists x \in t\, \varphi.$$

3. If φ and ψ are transitively bounded formulas, then also $\neg \varphi$, $(\varphi \wedge \psi)$, $(\varphi \vee \psi)$, $(\varphi \rightarrow \psi)$, $(\varphi \leftrightarrow \psi)$.

3.4.2 Superstructures

Definition 12. Let S be a nonempty set of atoms. For $n \in \mathbb{N}_0$ let S_n be defined inductively by $S_0 := S$ and $S_{n+1} := S_0 \cup \mathcal{P}(S_n)$. The set

$$\hat{S} := \bigcup \{S_n \mid n \in \mathbb{N}_0\}$$

is called the *superstructure of S*. The sets S_n are called the *levels* of $\hat{S}$.

The superstructure $\hat{S}$ and also each level S_n are transitive, i.e., for each set $A \in \hat{S}$ (or S_n), we also have $A \subseteq \hat{S}$ (or S_n). Furthermore,

$$S_0 \subseteq S_1 \subseteq \cdots \subseteq \hat{S} \quad \text{and} \quad S_0 \in S_1 \in \cdots \in \hat{S}.$$

We call $\hat{S}$ the *standard world*. In the following, a *nonstandard world* is placed next to it so that nonstandard methods are applicable.

In many applications, we find $S = \mathbb{R}$ or $S \supseteq \mathbb{R}$. However, S could also be any topological space, for example. Instead of $S = \mathbb{R}$, $S = \mathbb{N}_0$ would suffice in principle because the real numbers can then be constructed within $\hat{S}$ as usual. For finite S, $\hat{S}$ only contains finite sets. This case is not interesting for nonstandard analysis because the elementary embeddings defined below do not lead to a proper extension of the standard world.

3.4.3 Elementary embeddings

Let S and T be two nonempty sets of atoms and $\hat{S}$ and $\hat{T}$ their respective superstructures. The symbol set S should contain exactly one constant for each $s \in \hat{S}$ (and no other constants). Again, for simplicity, we assume that s itself is the constant for $s \in \hat{S}$. Further, let $\mathfrak{A} := (\hat{S}, \mathfrak{a})$ and $\mathfrak{B} := (\hat{T}, \mathfrak{b})$ be two S-structures[18] for which
1. $\mathfrak{a}(s) = s$ for all $s \in \hat{S}$,
2. $\mathfrak{a}(\in) = \in_{\hat{S}}$ and $\mathfrak{b}(\in) = \in_{\hat{T}}$.

The second condition states that the interpretation of the relation symbol $\in$ in the structures $\mathfrak{A}$ and $\mathfrak{B}$ are the $\in$-predicate of the background set theory (restricted to the domain of the respective structure). Thus the S-formula $a \in b$ (with constants a, b) is to be interpreted in $\mathfrak{A}$ as $a \in b$ and in $\mathfrak{B}$ as $\mathfrak{b}(a) \in \mathfrak{b}(b)$.

The convention of using the elements of $\hat{S}$ itself as constants of the formal language $\mathcal{L}^S$ has the advantage that an S-formula φ can be read directly as its interpretation in $\mathfrak{A}$ (if one replaces $\rightarrow$ by $\Rightarrow$ and $\leftrightarrow$ by $\Leftrightarrow$), although there is, of course, a formal difference. Example: In the S-formula $1 \in \{1, 2\}$, 1 and $\{1, 2\}$ are each *one* constant. The interpretation in $\mathfrak{A}$ looks the same, but is written in the language of background set theory (with the defined constants 1 and 2 and the defined operator symbol $\{.,.\}$).

Definition 13. The mapping $* : \hat{S} \rightarrow \hat{T}, s \mapsto {}^*s := \mathfrak{b}(\mathfrak{a}^{-1}(s))$ is called an *elementary embedding*, provided that
1. for each transitively bounded sentence φ, if $\mathfrak{A} \vDash \varphi$, then $\mathfrak{B} \vDash \varphi$.
2. $^*S = T$.

Because of the second condition, the embedding is also written $* : \hat{S} \rightarrow \widehat{{}^*S}$. Since with φ also $\neg\varphi$ is transitively bounded, the first condition also implies the inverse: If

---

18 Here $\mathfrak{a}, \mathfrak{b}$ denote the mappings that assign the symbols in S to their interpretations in $\mathfrak{A}$ and $\mathfrak{B}$, respectively.

$\mathcal{B} \models \varphi$, then $\mathfrak{A} \models \varphi$. If $\varphi^{\hat{S}}$ and $\varphi^{\hat{T}}$ denote the interpretations of φ in $\mathfrak{A}$ and $\mathcal{B}$ respectively, we obtain the following transfer principle:

Theorem 46 (transfer principle). *Let* $* : \hat{S} \to \hat{T}$ *be an elementary embedding and* φ *a transitively bounded sentence. Then*

$$\varphi^{\hat{S}} \Leftrightarrow \varphi^{\hat{T}}.$$

Various compatibility statements for the mapping $*$ can be derived from Theorem 46, for example (cf. [224], pp. 26 and 29–30):

1. For all $a, b \in \hat{S}$,

$$a = b \quad \Leftrightarrow \quad {}^*a = {}^*b, \tag{3.11}$$

$$a \in b \quad \Leftrightarrow \quad {}^*a \in {}^*b, \tag{3.12}$$

$$a \subseteq b \quad \Leftrightarrow \quad {}^*a \subseteq {}^*b. \tag{3.13}$$

2. For all $a \in \hat{S}$,

$$a \in S \quad \Leftrightarrow \quad {}^*a \in {}^*S. \tag{3.14}$$

3. For all $a_1, \ldots, a_n \in \hat{S}$,

$$*\{a_1, \ldots, a_n\} = \{{}^*a_1, \ldots, {}^*a_n\}, \tag{3.15}$$

$$*(a_1, \ldots, a_n) = ({}^*a_1, \ldots, {}^*a_n). \tag{3.16}$$

4. For all sets $A, B \in \hat{S}$,

$$*(A \cup B) = {}^*A \cup {}^*B, \tag{3.17}$$

$$*(A \cap B) = {}^*A \cap {}^*B, \tag{3.18}$$

$$*(A \setminus B) = {}^*A \setminus {}^*B, \tag{3.19}$$

$$*(A \times B) = {}^*A \times {}^*B. \tag{3.20}$$

In particular, $*\emptyset = \emptyset$.

5. For all binary relations $R \in \hat{S}$,
   (a) $\operatorname{dom}({}^*R) = {}^*\operatorname{dom}(R)$ and $\operatorname{ran}({}^*R) = {}^*\operatorname{ran}(R)$.
   (b) R is a function if and only if $*R$ is a function.

For given sets $A, B \in \hat{S}$, we generally do *not* have $*\mathcal{P}(A) = \mathcal{P}({}^*A)$ and $*(B^A) = {}^*B^{{}^*A}$ (see instead Theorem 49). **!**

According to (3.11), $*$ is injective and, according to (3.14), it maps atoms to atoms and sets to sets.

The designation $^*a$ for the image of a under the mapping $*$ has become established in nonstandard analysis. The attributive notation usually used for mappings would be misleading for arguments that are sets since $*(A)$ would stand for the image of the set A under the mapping $*$, but, according to common convention, also for the set of images of the elements of A. In nonstandard analysis, however, there is a decisive difference between these two sets. Väth designates the latter as $^\sigma A$, i. e.,

$$^\sigma A := \{^*a \mid a \in A\}. \tag{3.21}$$

For all sets $A \in \hat{S}$,

$$^\sigma A \subseteq {^*A} \tag{3.22}$$

(cf. [224], p. 26)

From (3.15), it follows that equality holds in (3.22) if A is finite. For infinite sets, however, the inclusion can be proper.

3.4.4 Nonstandard embeddings

The inclusion $^\sigma A = \{^*a \mid a \in A\} \subseteq {^*A}$ means that an elementary embedding $*$ applied to a set A takes the elements of A with it, but (for infinite sets) possibly adds further elements. This "blow-up"[19] is precisely the characteristic (and desired) effect of *nonstandard embeddings*.

Definition 14. An elementary embedding $* : \hat{S} \to \widehat{^*S}$ is called a *nonstandard embedding* provided $^\sigma A \neq {^*A}$ holds for all infinite sets $A \in \hat{S}$.

It turns out that an elementary embedding is already a nonstandard embedding if $^\sigma A \neq {^*A}$ holds for at least one infinite set $A \in \hat{S}$ (cf. [224], p. 40).

3.4.5 Standard elements and standard sets

The elements of $\operatorname{ran}(*)$ with an elementary embedding $* : \hat{S} \to \widehat{^*S}$ are called the *standard elements* of $\widehat{^*S}$. If these are sets, relations or functions, they are referred to as *standard sets, standard relations, standard functions*.

From the compatibility statements (3.17) to (3.20), it follows that with standard sets A, B also $A \cup B, A \cap B, A \setminus B, A \times B$ are standard sets. The following applies in general:

Theorem 47 (standard definition principle). *A set $A \in \widehat{^*S}$ is a standard set if and only if there exists a standard set B, a transitively bounded formula φ with the only free vari-*

---

19 Väth writes: "It is a good idea to think of $*$ as a 'blow-up-functor'" ([224], p. 24).

ables $x, x_1, \ldots, x_n$ ($n = 0$ not excluded) and standard elements $b_1, \ldots, b_n$ such that

$$A = \{b \in B \mid \varphi^{\widehat{*S}}[b, b_1, \ldots, b_n]\}$$

(cf. [224], S. 28).

In short, standard sets are obtained by separation from standard sets with transitively bounded standard formulas (i. e., S-formulas interpreted in $\widehat{*S}$ that may have standard elements as parameters).

3.4.6 Internal elements and internal sets

Definition 15. Let $* : \widehat{S} \to \widehat{*S}$ be an elementary embedding. The elements of the standard sets are called *internal* The set $\mathcal{I}$ of all internal elements of $\widehat{*S}$ is called the *nonstandard world*, i. e.,

$$\mathcal{I} := \bigcup \{{}^*A \mid A \in \widehat{S} \setminus S\}.$$

All elements of $\widehat{*S}$ that are not internal are called *external*.

If the elements of $\widehat{*S}$ are sets, relations, functions, one speaks accordingly of *internal* or *external sets, relations, functions*.

Further, $\mathcal{I}$ is a transitive subset of $\widehat{*S}$, and

$$\mathcal{I} = \bigcup_{n=0}^{\infty} {}^*S_n$$

(cf. [224], S. 36).

Theorem 48 (internal definition principle). *A set $A \in \widehat{*S}$ is internal if and only if there exists an internal set B, a transitively bounded formula φ with the only free variables $x, x_1, \ldots, x_n$ ($n = 0$ not excluded) and internal elements $b_1, \ldots, b_n$ such that*

$$A = \{b \in B \mid \varphi^{\widehat{*S}}[b, b_1, \ldots, b_n]\}$$

(cf. [224], S. 37).

In short, internal sets are obtained by separation from internal sets with transitively bounded internal formulas (i. e., S-formulas interpreted in $\widehat{*S}$ that may have internal elements as parameters).

From the internal definition principle follow analogous compatibility statements as for standard sets, in particular that for internal sets A, B also $A \cup B, A \cap B, A \setminus B, A \times B$ are internal sets.

Theorem 49 ([224], p. 40). *Let* $* : \hat{S} \to \widehat{{}^*S}$ *be an elementary embedding. Then for all sets* $A \in \hat{S}$,

$$^*\mathcal{P}(A) = \{M \subseteq {}^*A \mid M \text{ is internal}\}$$

and for all sets $A, B \in \hat{S}$,

$$^*(B^A) = \{f \in {}^*B^{*A} \mid f \text{ is internal}\}.$$

A statement that is true in the standard world for all subsets of a set A or for all functions from A to B is true in the nonstandard world for all *internal* subsets of A or for all *internal* functions from A to B because sentences of the form $\forall X \in \mathcal{P}(A)\ldots$ or $\forall f \in B^A \ldots$ are transitively bounded (with the constants $\mathcal{P}(A)$ or B^A) and can therefore be transferred to the nonstandard world using the transfer principle.

In the nonstandard embedding $\hat{\mathbb{R}} \to \widehat{{}^*\mathbb{R}}$, for example, the well-ordering principle applies to ${}^*\mathbb{N}$ and the least-upper-bound property applies to ${}^*\mathbb{R}$, each restricted to *internal* subsets:

- Every nonempty *internal* subset of ${}^*\mathbb{N}$ contains a smallest number.
- Every nonempty *internal* subset of ${}^*\mathbb{R}$ that is bounded above has a least upper bound in ${}^*\mathbb{R}$.

The embedded sets ${}^\sigma\mathbb{N}, {}^\sigma\mathbb{R}$ and their complements in ${}^*\mathbb{N}$ and ${}^*\mathbb{R}$ are external. The above statements do not apply to them. For example, the nonempty subset ${}^*\mathbb{N} \setminus {}^\sigma\mathbb{N} \subseteq {}^*\mathbb{N}$ does not contain a smallest number, and the nonempty set ${}^\sigma\mathbb{R} \subseteq {}^*\mathbb{R}$ is bounded above, but has no least upper bound in ${}^*\mathbb{R}$.

In general, for any nonstandard embedding, ${}^\sigma A$ is external for all infinite sets A (see [224], p. 41).

3.4.7 Hyperinfinite sets

Definition 16. A set $A \in \widehat{{}^*S}$ is called **-finite* or *hyperfinite* if there is an internal bijective mapping $f : \{k \in {}^*\mathbb{N}_0 \mid 1 \leq k \leq n\} \to A$ with $n \in {}^*\mathbb{N}_0$. In this case, one defines ${}^\#A := n$, otherwise one writes ${}^\#A = \infty$.

The number ${}^\#A$ is well defined because for a hyperfinite set A the $n \in {}^*\mathbb{N}_0$ in Definition 16 is uniquely determined (see [224], p. 78). Instead of $\{k \in {}^*\mathbb{N}_0 \mid 1 \leq k \leq n\}$, one also writes more suggestively $\{1, \ldots, n\}$. Note, however, that this set is uncountable in the case $n \gg 1$.

The hyperfinite sets behave formally like finite sets. For example, every hyperfinite set of hyperreal numbers always has a smallest and a largest element, and the following applies to any hyperfinite sets A, B (cf. [224], p. 82):

- ${}^\#(A \times B) = {}^\#A \cdot {}^\#B$,

- $^\#(A \cup B) = {}^\#A + {}^\#B$, if $A \cap B = \emptyset$,
- $^\#(^*\mathcal{P}(A)) = 2^{\#A}$.

Furthermore, hyperfinite sums and products of hyperreal numbers can be defined. Let $\mathbb{R}^{<\mathbb{N}}$ be the set of all finite sequences in $\mathbb{R}$ and $\Sigma\colon \mathbb{R}^{<\mathbb{N}} \to \mathbb{R}$ the ordinary sum function. Then $^*(\mathbb{R}^{<\mathbb{N}})$ contains all hyperfinite sequences $f\colon \{1,\ldots,h\} \to {}^*\mathbb{R}$, $h \in {}^*\mathbb{N}$, and one defines $\sum_{n=1}^{h} f(n) := {}^*\Sigma(f)$ (cf. [224], p. 84). Proceed analogously for products. Hyperfinite sums play a decisive role for more general integral definitions (cf. [127], pp. 156–171). The Riemann integral can be defined as in Section 3.2.5 as the standard part of hyperfinite Riemann sums.

3.4.8 Analysis for internal functions

The possibility of using calculus with nonstandard functions is a real enrichment compared to conventional analysis. Schmieden and Laugwitz have already shown in [190] how Dirac's so-called "delta functions" can be realized as *nonstandard functions*, which is simpler and more flexible than the otherwise necessary delta distributions.[20] The idea with a delta function is that it should vanish everywhere except at point 0, but have the integral 1. This is not possible for ordinary real functions. A delta function can be approximated by a function of the form

$$\delta_n(x) := \frac{n}{\pi(1 + x^2 n^2)}$$

with very large n. This function is continuous everywhere and has the integral 1, as desired. The larger the n, the more the area under the graph is concentrated at 0.

In nonstandard analysis, we can choose $n \gg 1$ and thus obtain a function that has infinitesimal values for noninfinitesimal x and the infinite value n at point 0. But how should we define its integral?

It is immediately obvious that a function that grows from infinitesimal to noninfinitesimal or even infinite values within an infinitesimal interval cannot be continuous in the sense that $x \approx x_0$ always implies $f(x) \approx f(x_0)$. Likewise, it cannot be expected that the differential quotient at a point x_0 is independent of dx or that the Riemann sum between two points is independent of the chosen infinitely fine partition. The question therefore arises as to how concepts of real analysis (we call them $\mathbb{R}$-concepts for short), such as limits, continuity, derivative, integral, can be generalized to internal hyperreal functions.

One obvious possibility is to transfer the classical ε-definitions to the hyperreal case. This leads, for example, to the following definitions:

---

[20] Before Schmieden and Laugwitz, Cauchy had already used the language of infinitesimals in 1815 to describe a function that has the properties of the delta function (see [46], p. 289).

- $a \in {}^*\mathbb{R}$ is called the *limit* of the internal hyperreal sequence (a_n), if for every hyperreal $\varepsilon > 0$ there is $n_0 \in {}^*\mathbb{N}$ such that for all $n \in {}^*\mathbb{N}$,

$$n \geq n_0 \quad \Rightarrow \quad |a_n - a| < \varepsilon.$$

- An internal hyperreal function f is said to be *continuous* at x_0 if $x_0 \in \text{dom}(f)$ and if for every hyperreal $\varepsilon > 0$ there is a hyperreal $\delta > 0$ such that for all $x \in \text{dom}(f)$,

$$|x - x_0| < \delta \quad \Rightarrow \quad |f(x) - f(x_0)| < \varepsilon.$$

- $m \in {}^*\mathbb{R}$ is called the *derivative* of the internal hyperreal function f at x if x is an accumulation point of $\text{dom}(f)$ (i. e., if every ε-neighborhood of x, with $\varepsilon \in {}^*\mathbb{R}$, $\varepsilon > 0$, contains an element of $\text{dom}(f) \setminus \{x\}$) and if for every hyperreal $\varepsilon > 0$ there is a hyperreal $\delta > 0$, so that for all $h \in {}^*\mathbb{R} \setminus \{0\}$ with $x + h \in \text{dom}(f)$ and $|h| < \delta$,

$$\left| \frac{f(x+h) - f(x)}{h} - m \right| < \varepsilon.$$

- $s \in {}^*\mathbb{R}$ is the *integral* of the internal hyperreal function f from a to b if $[a, b] \subseteq \text{dom}(f)$ is a nondegenerate hyperreal interval and if for every hyperreal $\varepsilon > 0$ there exists a hyperreal $\delta > 0$, so that for every Riemann sum of mesh $< \delta$, $|r - s| < \varepsilon$ (where r is the value of the Riemann sum).

As in real analysis, the limit, derivative, and integral are well defined (i. e., unique if they exist). The terms *convergent*, *differentiable*, *integrable* are defined as usual via the existence of the limit, the derivative, or the integral. All classical theorems of analysis, for example, the intermediate value theorem, the mean value theorems or the fundamental theorem of differential and integral calculus, can be directly transferred to the hyperreal case for internal functions using the transfer principle. In particular, the function $\delta_n(x)$ defined above, with $n \gg 1$, is internal (by the internal definition principle), continuous everywhere, and has the integral 1.

3.4.9 Implicit definitions via *

Another way to generalize the $\mathbb{R}$-concepts to internal hyperreal functions is to apply the mapping * to a set defining the respective $\mathbb{R}$-concepts and take advantage of the fact that in this way the internal nonstandard elements are supplied by the "blow-up" effect. No recourse to the ε-definitions is then necessary.

In this section, let $\mathcal{F}$ be the set of all real-valued functions with $\text{dom}(f) \subseteq \mathbb{R}$. Suppose, the $\mathbb{R}$-concepts limit, continuity, derivative, integral have been defined by the nonstandard criteria from Definition 10 in Section 3.2.5 for real-valued functions. The generalization to internal functions can then be carried out as follows:

- A number $a \in {}^*\mathbb{R}$ is called *limit* of the internal hyperreal sequence $(f(n))_{n \in {}^*\mathbb{N}}$ if $(f, a) \in \mathcal{E}_1$, where

$$\mathcal{E}_1 := {}^*\left\{(f, a) \in \mathbb{R}^{\mathbb{N}} \times \mathbb{R} \mid \lim_{n \to \infty} f(n) = a\right\}.$$

- An internal hyperreal function f is said to be *continuous* at x_0 if $(f, x_0) \in \mathcal{E}_2$, where

$$\mathcal{E}_2 := {}^*\{(f, x_0) \in \mathcal{F} \times \mathbb{R} \mid f \text{ continuous at } x_0\}.$$

- A number $m \in {}^*\mathbb{R}$ is called *derivative* of the internal hyperreal function f at x if $(f, x, m) \in \mathcal{E}_3$, where

$$\mathcal{E}_3 := {}^*\{(f, x, m) \in \mathcal{F} \times \mathbb{R} \times \mathbb{R} \mid f'(x) = m\}.$$

- A number $s \in {}^*\mathbb{R}$ is called *integral* of the internal hyperreal function f from a to b if: $(f, a, b, s) \in \mathcal{E}_4$, where

$$\mathcal{E}_4 := {}^*\left\{(f, a, b, s) \in \mathcal{F} \times \mathbb{R} \times \mathbb{R} \times \mathbb{R} \,\Big|\, \int_a^b f(x)\, dx = s\right\}.$$

These definitions (called *-definitions for short to distinguish them from the ε-definitions given above) can also be used to transfer all theorems of classical analysis to the internal hyperreal case using the transfer principle. In particular, the *-definitions are equivalent to the ε-definitions.

Sometimes new names are given to the generalized concepts, such as *-continuity, *-derivative, *-integral (see, for example, [127]). However, no misunderstandings are to be expected if the old terms continue to be used. This follows the previous practice of omitting the asterisk in the designation when extending functions and relations.

3.4.10 Enlargements and saturation

For infinitely fine partitions of an interval $[a, b]$, as they occur in the nonstandard definition of the integral, the uncountable real numbers of the interval are isolated from each other by n subintervals ($n \in {}^*\mathbb{N}, n \gg 1$). The set ${}^*\mathbb{N}$ must therefore have at least the cardinality of the real interval and thus the cardinality of $\mathbb{R}$. On the other hand, it follows from the calculation rules for cardinal numbers that with the construction from Section 3.3.1 ${}^*\mathbb{R}$ cannot have a greater cardinality than $\mathbb{R}$. Therefore, ${}^*\mathbb{N}$, ${}^*\mathbb{R}$, and $\mathbb{R}$ are equipotent.

It is plausible that this is no longer sufficient for advanced applications in topology, functional analysis, or stochastics. To be able to embed every set A of the superstructure into a hyperfinite set, there must be sufficiently many hypernatural numbers. This requirement leads to the notion of an *enlargement*.

We say that a family $\mathcal{A}$ of sets has the *finite intersection property* if for each collection of finitely many sets of $\mathcal{A}$ the intersection is nonempty.

Definition 17 ([224], p. 103). Let $* : \hat{S} \to {}^*\widehat{S}$ be an elementary embedding. Then $*$ is called an *enlargement* if for every $\mathcal{A} \subseteq \hat{S} \setminus S$ with the finite intersection property, $\bigcap {}^\sigma \mathcal{A}$ is nonempty, i. e.,

$$\bigcap \{{}^*A \mid A \in \mathcal{A}\} \neq \emptyset.$$

Definition 18 ([224], p. 107). Let $R \in \hat{S}$ be a binary relation and $A \subseteq \mathrm{dom}(R)$. We say that R is *satisfied by* $b \in \mathrm{ran}(R)$ *on* A if $(a, b) \in R$ for each $a \in A$. We call R *concurrent* or *finitely satisfiable* on A if for each finite subset $A_0 \subseteq A$ there is some $b \in \mathrm{ran}(R)$ which satisfies R on A_0.

The following theorem shows how enlargements are related to concurrent relations and hyperfinite sets.

Theorem 50 ([224], p. 107). *Let* $* : \hat{S} \to {}^*\widehat{S}$ *be an elementary embedding. Then the following statements are equivalent:*
1. *$*$ is an enlargement.*
2. *For every binary relation $R \in \hat{S}$, if R is concurrent on $\mathrm{dom}(R)$, then there is some $b \in \mathrm{ran}({}^*R)$ which satisfies ${}^*R$ on ${}^\sigma \mathrm{dom}(R)$, i. e., for which*

$$({}^*a, b) \in {}^*R \quad \text{for all } a \in \mathrm{dom}(R).$$

3. *For every set $A \in \hat{S}$, there is a hyperfinite set H with*

$$ {}^\sigma A \subseteq H \subseteq {}^*A, $$

where ${}^\sigma A = \{{}^*a \mid a \in A\}$.

So with an enlargement, any set of the standard world can be embedded in a hyperinfinite set of the nonstandard world. Every enlargement is a nonstandard embedding ([224], p. 105).

Some applications of nonstandard analysis require nonstandard embeddings with a property called κ-saturation (where κ is a sufficiently large cardinal).

Definition 19 ([56], p. 44). Let κ be a cardinal number. A nonstandard embedding is called κ-*saturated* if for any family $\mathcal{B}$ of internal sets which has the finite intersection property and cardinality strictly less than κ, the total intersection $\bigcap \mathcal{B}$ is nonempty.[21]

---

21 Väth uses a slightly different terminology, in which κ may be any set (not necessarily a cardinal) and $\mathcal{B}$ is required to have at most the cardinality of κ. For example, he says $\mathbb{N}$-saturated instead of $\aleph_1$-saturated where $\aleph_1$ is the least uncountable cardinal (see his warning on p. 103 in [224]).

For $\kappa > |\widehat{S}|$, the embedding is also called *polysaturated*.[22] Polysaturated nonstandard embeddings are always also enlargements ([224], p. 104).

3.4.11 The existence of nonstandard embeddings

A proof for the existence of nonstandard embeddings can be found, for example, in [224], pp. 51–56. The proof employs δ-incomplete ultrafilters on an index set J. A filter $\mathcal{F}$ is called δ-incomplete if there is a countable subset $\mathcal{F}_0 \subseteq \mathcal{F}$ with $\bigcap \mathcal{F}_0 \notin \mathcal{F}$.

For ultrafilters on a countable index set J (for example, $J = \mathbb{N}$), the properties *free* and *δ-incomplete* are equivalent (see [224], pp. 47–48).

Ultrafilters on uncountable index sets J can be used, for example, to prove the existence of enlargements and polysaturated nonstandard embeddings (see, for example, [127], pp. 410–428).[23]

3.5 Internal set theory

Having considered nonstandard extensions of superstructures, one might ask whether the entire set universe can be extended in a nonstandard way. This is the motivation behind axiomatic approaches such as Edward Nelson's *internal set theory*.

The basic idea is to enrich the language of set theory with a new undefined predicate called "standard" and to postulate (by suitable axioms) that this applies to all classically definable sets (i.e., sets definable by ϵ-formulas), but not to all sets. This gives the set universe an additional quality that is invisible to classical set theory. This is sometimes compared to a transition from seeing in black and white to seeing in color (see, e.g., [159]). We have already had a first look at internal set theory in Chapter 2, since the theory used there is a weakened version of IST, essentially SPOT (see Sections 2.10 and 2.11). The following presentation is based on [160].

3.5.1 Extension of the language

The symbol set of the ZFC set theory contains only one symbol, the binary relation symbol ϵ. This is not defined, but is, so to speak, part of the basic equipment of the language of set theory. All other symbols commonly used in set theory are defined in ZFC by ϵ-formulas. They are *defined symbols*.

---

[22] In Väth's terminology, *polysaturated* is the same as $\widehat{S}$-*saturated*.
[23] I am referencing Landers and Rogge here because Väth uses a different method to construct the existence of strong and $\widehat{S}$-compact nonstandard embeddings.

In Nelson's *Internal Set Theory* (IST), a new *undefined symbol*, the unary relation symbol st (for the unary predicate "standard"), is added to the symbol set of ZFC. The axiom system ZFC is supplemented by three additional axiom schemas that specify how to handle the new predicate st.

Formulas that contain the symbol st directly or indirectly (via defined symbols) are called *external*, all other formulas are called *internal*. The internal formulas are therefore exactly those that can also be formulated in the original language of ZFC.

The axioms of ZFC are adopted unchanged in IST. In particular, this means that the axiom schemas of separation and replacement only apply to ∈-formulas, i.e., only to internal formulas, with the initially strange consequence that external formulas are generally not set-forming, i.e., that for any set A and an external formula φ it is generally not possible to conclude the existence of the set $\{x \in A \mid \varphi\}$. In particular, it is generally not possible to separate the standard elements of a set, i.e., to form the set $\{x \in A \mid \text{st}(x)\}$. Nelson calls a separation with external predicates an *illegal set formation* ([160], p. 1165). We had already encountered this special feature in Chapter 2.

On the other hand, the unchanged adoption of the ZFC axioms means that the entire stock of familiar ZFC mathematics is retained. All definitions remain unchanged, for example, the definitions of the sets $\mathbb{N}, \mathbb{Z}, \mathbb{Q}, \mathbb{R}, \mathbb{C}$ or the definition of the predicate *finite*. All theorems proved in ZFC remain valid without any change, for example the well-ordering principle for $\mathbb{N}$, the least-upper-bound property of $\mathbb{R}$, or the theorem that subsets of finite sets are finite again.

Furthermore, it can be shown that IST is a *conservative extension* of ZFC (cf. [160], pp. 1192–1197).[24] This means that every internal theorem that is provable in IST is already provable in ZFC. IST can therefore be seen as an optional additional tool for classical mathematics, whereby the classical results that can be obtained with it do not depend on additional assumptions. This distinguishes IST from ZFC extensions which, for example, require the existence of inaccessible cardinals. In particular, it follows from the conservativeness that IST is consistent relative to ZFC.

The additional axiom schemas in IST are called *Idealization* (I), *Standardization* (S), and *Transfer* (T). Thus, the acronym IST can also be interpreted as an abbreviation for the additional axiom schemas.

It is common (cf. [160] or [178]) to use the logical symbols of background set theory in IST, i.e., $\Rightarrow, \Leftrightarrow$ instead of $\rightarrow, \leftrightarrow$. Quantifiers relativized by st are also used to formulate the axioms. So $\forall^{\text{st}} x\, \varphi$ stands for $\forall x\, (\text{st}(x) \Rightarrow \varphi)$ and $\exists^{\text{st}} x\, \varphi$ for $\exists x\, (\text{st}(x) \wedge \varphi)$.

Let fin(z) be the abbreviation for a common formalization of the property "z is finite" in ZFC (for example, a formalization of: "Every injective mapping from z to z is also surjective"). Thus fin(z) is an internal formula. Accordingly, $\forall^{\text{st fin}} x\, \varphi$ stands for $\forall^{\text{st}} x\, (\text{fin}(x) \Rightarrow \varphi)$ and $\exists^{\text{st fin}} x\, \varphi$ for $\exists^{\text{st}} x\, (\text{fin}(x) \wedge \varphi)$.

---

24 Nelson states that this result goes back to William C. Powell.

3.5.2 Transfer

The axiom schema of *transfer* expresses that an internal property that applies to all standard sets already applies to all sets. It also ensures that all sets that can be defined in ZFC are standard. It corresponds to the transfer axiom from Section 2.6.1.

Axiom schema (transfer). *Let $\varphi(x, t_1, \ldots, t_k)$ be an internal formula with free variables $x, t_1, \ldots, t_k$ and no other free variables. Then*

$$\forall^{st} t_1 \ldots \forall^{st} t_k \left(\forall^{st} x \, \varphi(x, t_1, \ldots, t_k) \Rightarrow \forall x \, \varphi(x, t_1, \ldots, t_k) \right). \tag{T}$$

In this context, $t_1, \ldots, t_k$ are called *standard parameters* of φ, as the range of these variables is restricted to standard objects. Trivially, the direction "$\Leftarrow$" also holds in (T).

As in Section 2.6.1, it follows from (T) that two standard sets are equal if they contain the same standard elements (see Theorem 7). A standard set is therefore already uniquely determined by its standard elements.

Applying (T) to $\neg \varphi$ produces the dual variant (T'):

$$\forall^{st} t_1 \ldots \forall^{st} t_k \left(\exists x \, \varphi(x, t_1, \ldots, t_k) \Rightarrow \exists^{st} x \, \varphi(x, t_1, \ldots, t_k) \right). \tag{T'}$$

The opposite direction is again trivial.

In particular, it follows from (T'): If there is a unique x with $\varphi(x, t_1, \ldots, t_k)$, then this x must be standard. Therefore, everything that can be defined by internal formulas (possibly with standard parameters) is standard (see the examples in Section 2.6.1).

For an internal formula φ, let φ^{st} be the formula obtained by replacing all occurring quantifiers $\forall$ and $\exists$ by the relativized quantifiers $\forall^{st}$ and $\exists^{st}$, respectively. One calls φ^{st} the *relativization of φ to standard sets*. By repeated application of (T) or (T') (working from the outside in), we get

Theorem 51 ([160], p. 1166). *For each internal formula $\varphi(t_1, \ldots, t_n)$, we have*

$$\forall^{st} t_1 \ldots \forall^{st} t_n \left(\varphi^{st}(t_1, \ldots, t_n) \Leftrightarrow \varphi(t_1, \ldots, t_n) \right). \tag{3.23}$$

In particular, for each internal sentence φ (i. e., in the case $n = 0$),

$$\varphi^{st} \Leftrightarrow \varphi.$$

The transfer axiom is given in this form, for example, in [110], p. 84.

In order to be able to apply transfer to a statement, we must first verify two things ([160], p. 1166):
1. The statement is internal.
2. All parameters have standard values.

Nelson calls the violation of this rule an *illegal transfer* (ibid.). We had given a corresponding hint in Section 2.6.1.

In practice, it is often the case that the internal statement to be transferred contains defined constants that are known to be standard. In this case, the transfer is legal because (T) is applicable to formulas with standard parameters.

3.5.3 Idealization

The axiom schema of *idealization* ensures that there are nonstandard sets at all and that sufficiently large finite sets are available for all intended applications. It is much more general than the idealization axiom for real numbers in Section 2.6.2.

Axiom schema (idealization). *Let $\varphi(x, y)$ be an internal formula with free variables x and y (and possibly other free variables). Then*

$$\forall^{\text{st fin}} z \, \exists x \, \forall y \in z \, \varphi(x, y) \quad \Leftrightarrow \quad \exists x \forall^{\text{st}} y \, \varphi(x, y). \tag{I}$$

Using the phrase "x dominates y" for $\varphi(x, y)$ (imagine, for example, the relation $x > y$ on $\mathbb{R}$), the idealization axiom says that the following statements are equivalent:
1. For every finite standard set z there is an x that dominates all $y \in z$.
2. There is an x that dominates all standard y.

For most applications of (I) the direction $1 \Rightarrow 2$ is relevant. The reverse direction is used for the following characterization of finite standard sets

Theorem 52 ([160], p. 1167). *Let A be a set. Then each element of A is standard if and only if A is a finite standard set.*

From Theorem 52 it follows that every infinite set contains nonstandard elements.

In Section 3.4 the term *enlargement* was introduced, which allowed every set of the standard world to be embedded in a hyperfinite set of the nonstandard world. A characteristic feature of an enlargement was that every concurrent relation of the standard world is satisfiable in the nonstandard world on its entire domain (see Theorem 50). Axiom (I) (in the direction "$\Rightarrow$") is the axiomatic version of this characterization, now for predicates defined on the entire set universe.[25] The finite sets of the standard world $\hat{S}$ correspond to the finite standard sets in IST, the hyperfinite sets of the nonstandard world correspond to the finite sets in IST. Accordingly, it follows from (I) that there are "very large" finite sets.

Theorem 53. *There is a finite set that contains all standard sets as elements.*

---

[25] The position of the two parameters of φ is swapped in (I) compared to Definition 18, but this is only a matter of convention.

To prove this, apply (I) to the internal formula $(\mathrm{fin}(x) \wedge y \in x)$ (see [160], p. 1167).

Another important conclusion from (I) is the existence of "infinitely large" numbers in $\mathbb{N}$ (and thus also in $\mathbb{R}$).

Theorem 54 ([127], p. 438). *There is an $h \in \mathbb{N}$ with $h > n$ for all standard $n \in \mathbb{N}$.*

To prove this, apply (I) to the internal formula $(x \in \mathbb{N} \wedge (y \in \mathbb{N} \Rightarrow x > y))$. Since every number that can be represented by a sum $1 + \cdots + 1$ is standard, h is greater than each of these numbers and therefore greater than $1, 2, 3, \ldots$ The natural number h is therefore "infinitely large" in this sense. As explained in Section 2.6.2, the term *unlimited* is preferred in axiomatic nonstandard theories.

In $\mathbb{N}$ all standard numbers come *before* all nonstandard numbers because for each standard $n \in \mathbb{N}$ the finite set $\{k \in \mathbb{N} \mid k < n\}$ is standard and therefore contains only standard elements according to Theorem 52. In contrast, compare the proof of Theorem 13 in Section 2.6.3, which gets by with the much weaker idealization axiom for real numbers (and the standard part axiom).

3.5.4 Standardization

The axiom schema of standardization is a certain compensation for the inapplicability of the separation axiom for external formulas. We have already discussed it in Section 2.10.1 and give here again its formulation in the context of internal set theory.

Axiom schema (standardization). *Let $\varphi(z)$ be an (internal or external) formula with the free variable z (and possibly other free variables). Then*

$$\forall^{st} x \exists^{st} y \forall^{st} z \, (z \in y \Leftrightarrow z \in x \wedge \varphi(z)). \tag{S}$$

We note once again that (S) has the same structure as the axiom of separation, apart from the fact that all quantifiers are relativized by st.

Colloquially, (S) says: For every standard set x, there is a standard set y whose standard elements are precisely the standard elements of x that satisfy φ. The standard set y is thus uniquely determined and a subset of x. It is denoted by $^S\{z \in x \mid \varphi(z)\}$.

The following theorem shows that an external predicate can be used to define a standard function as long as standard sets are assigned to standard sets. For example, one can define the derivative function of a standard function whose derivative is defined (by an external predicate) at each standard point of its domain.

Theorem 55. *Let X and Y be standard sets and $\varphi(x,y)$ a binary predicate such that for every standard $x \in X$ there is a unique standard $y \in Y$ with $\varphi(x,y)$. Then there is a unique standard function $f: X \to Y$ such that for all standard $x \in X$ and for all standard $y \in Y$,*

$$f(x) = y \quad \Leftrightarrow \quad \varphi(x,y). \tag{3.24}$$

Proof. Define $f := {}^S\{(x,y) \in X \times Y \mid \varphi(x,y)\}$. By definition, the standard elements of f are precisely the standard $(x,y) \in X \times Y$ that satisfy $\varphi(x,y)$ (where (x,y) is standard if and only if x and y are standard). By hypothesis, for every standard $x \in X$ there is a unique standard $y \in Y$ with $\varphi(x,y)$. Therefore, for every standard $x \in X$ there is a unique standard $y \in Y$ with $(x,y) \in f$. This is an internal statement with standard parameter f. By transfer we get: For every $x \in X$, there is a unique $y \in Y$ with $(x,y) \in f$. Thus f is a function from X to Y and (by construction) satisfies (3.24) for all standard x, y. Since f is a standard set, it is uniquely determined by its standard elements.[26] □

3.5.5 Elementary analysis in internal set theory

In Robinson's nonstandard analysis, the real numbers are the standard numbers and the numbers added by field extension $^*\mathbb{R} \supset \mathbb{R}$ are the nonstandard numbers. In internal set theory, the distinction between standard and nonstandard numbers is made within the real numbers by the new predicate *standard*. Accordingly, the nonstandard definitions of concepts in calculus will differ from the definitions in Robinson's nonstandard analysis in that the standard numbers take on the role of the real numbers and the real numbers the role of the hyperreal numbers.

Instead of the necessity to generalize concepts such as *continuity, derivative*, etc., from the case of real numbers and functions to the case of hyperreal numbers and internal functions, in IST the necessity arises to generalize from standard numbers and functions to real numbers and functions. In Section 2.10, we have already seen how this can be achieved by means of standardization. The internal sets of Robinson's nonstandard analysis simply correspond to the sets in IST. The external sets of Robinson's nonstandard analysis have no equivalent in IST.[27]

The sets $\mathbb{N}$ and $\mathbb{R}$ are defined in IST as in ZFC, and they retain all their properties provable in ZFC. This means, for example, that $\mathbb{R}$ in IST has the least-upper-bound property. The terms *limited, unlimited* and *infinitesimal* are defined with the predicate *standard* as in Section 2.6.2 (see Definition 4 there). According to Theorem 54, $\mathbb{N}$ contains unlimited numbers. Their reciprocals are infinitesimal. Thus $\mathbb{R}$ contains both unlimited numbers and nonzero infinitesimals. The existence and uniqueness of the standard part of limited numbers now follows from (S).[28]

---

26 In [160], p. 1168, the theorem is proved in a more general version that needs the axiom of choice.

27 Since there are advanced applications of nonstandard analysis (e. g., nonstandard hulls or Loeb measures) that make use of external sets (in the model-theoretic approach), one might think that these applications are not accessible to internal set theories. However, this is generally not true (see [103]).

28 To prove the standard part principle, a weaker, countable version of standardization is sufficient (see Lemma 2.4 in [104]).

Theorem 56. *Every limited $x \in \mathbb{R}$ is infinitely close to a unique standard real number* stp(x) *(the standard part of x).*[29]

We quote the proof from [160], p. 1169.

Proof. Let x be a limited real number, so that there is a standard real number r with $|x| \leq r$. Let $E := {}^S\{t \in \mathbb{R} \mid t \leq x\}$. Then E is by definition a standard set, and for all standard $t \in E$ we have $t \leq r$. By transfer, for all $t \in E$ we have $t \leq r$. Therefore E is bounded above. The set E is nonempty since $-r \in E$. Therefore E has a least upper bound a. Since E is standard, a is standard, by transfer. Suppose that $x - a > \varepsilon$ for some standard $\varepsilon > 0$. Then $a + \varepsilon \leq x$, and since $a + \varepsilon$ is standard, $a + \varepsilon$ is in E, which contradicts the fact that a is an upper bound for E. Suppose that $a - x > \varepsilon$ for some standard $\varepsilon > 0$. Then $x \leq a - \varepsilon$, so that for any standard $t \in E$ we have $t \leq a - \varepsilon$ and by transfer (since E and $a - \varepsilon$ are standard) for all $t \in E$ we have $t \leq a - \varepsilon$, which contradicts the fact that a is the least upper bound of E. Consequently, $x \approx a$. The uniqueness is clear from the fact that 0 is the only standard infinitesimal. □

The central concepts of elementary analysis can then be defined classically or as in Section 2.10.2 and theorems can be proved as in Section 2.9.

3.6 Other axiomatic approaches

3.6.1 Bounded set theory

A variant of internal set theory is *Bounded Set Theory* (BST for short), which was introduced by Kanovei ([112], p. 16).

BST includes the ZFC axioms,[30] as well as the axiom schemas *transfer* and *standardization* (as in IST). In addition, there is the axiom of *boundedness* (which states that all sets are elements of some standard set) and the axiom schema of *bounded idealization* (instead of the idealization schema in IST).

Axiom (boundedness).
$$\forall x \exists^{st} y \, x \in y. \tag{3.25}$$

Axiom schema (bounded idealization).
$$\forall^{st} u \; [\forall^{st\,fin} z \subseteq u \, \exists x \, \forall y \in z \, \varphi(x,y) \quad \Leftrightarrow \quad \exists x \, \forall^{st} y \in u \, \varphi(x,y)] \tag{3.26}$$

---

[29] Sometimes the standard part of x is called the *shadow* of x and various designations for it can be found in the literature, for example, st x [160], sh(x) [104], x^* [178] or $°x$ [61].

[30] In other sources [110, 105], the ZFC axioms relativized to standard sets are included in BST instead of the original ZFC axioms. However, this is irrelevant since every ZFC axiom φ is an internal sentence and therefore we have $\varphi^{st} \Leftrightarrow \varphi$ by transfer (cf. Theorem 51).

for each internal formula $\varphi(x,y)$ with free variables x and y (and possibly other free variables).

The difference to the idealization schema (I) in IST is that the idealization (expressed by the equivalence) does not have to hold for the whole universe, but only within any standard set u. Accordingly, Theorem 53 only applies in a form bounded by standard sets: For every standard set u, there is a finite set containing all standard elements of u, which is usually sufficient for practical purposes. For the following statements on BST, see [110], pp. 131–137.

Like IST, BST is also a conservative extension of ZFC (i. e., every $\in$-sentence that is provable in BST is already provable in ZFC). The equiconsistency of BST and ZFC follows from the conservativeness.

In addition, BST has another property, which Kanovei calls *standard core interpretability* and which, among other things, ensures that any model of ZFC can be extended to a model of BST.

The standard core interpretability of BST means: There is an interpretation[31] of BST in ZFC such that (under this interpretation) the ZFC universe $\mathbf{V}$ is the class of standard sets in BST (the *standard core*). More precisely: There is a $\{\in, \text{st}\}$-structure $^*\mathbf{v} = (^*\mathbf{V}, ^*\in, ^*\text{st})$ and a $\in$-embedding $* : \mathbf{V} \to {^*\mathbf{V}}$ (an injective mapping, with $^*x \,{^*\in}\, ^*y \Leftrightarrow x \in y$ for all $x, y \in \mathbf{V}$) such that

$$\mathbb{S}^{(^*\mathbf{v})} := \{z \in {^*\mathbf{V}} \mid {^*\text{st}}(z)\} = \{^*x \mid x \in \mathbf{V}\}.$$

The class $\mathbb{S}^{(^*\mathbf{v})}$ is the *standard core* of $^*\mathbf{v}$. Informally speaking, we describe an extension of the ZFC universe within ZFC in such a way that the standard core of the extension is the initial ZFC universe. Kanovei calls standard core interpretable theories "realistic" (whereby he himself always puts the term in quotation marks).

In contrast to BST, IST is not standard core interpretable. There are models of ZFC that cannot be extended to a model of IST. A detailed discussion of the advantages of BST over IST can be found in [110].

3.6.2 External set theories

In internal set theories such as IST or BST, external predicates generally are not set forming. Therefore, for example, the standard elements of $\mathbb{N}$ or $\mathbb{R}$ do not form a set. External set theories also allow the formation of such *external sets* and are closer to the model-theoretic approach in this respect. However, they are more complicated to describe and have other restrictions regarding set formation, for example, in such a way that the

---

31 The terms *interpretation, structure, mapping, injective*, etc., are to be understood in this context in relation to *classes*. For the definition of the class $\mathbf{V}$, see also Section 5.4.2.

power set of a set or the set of all mappings from one set to another cannot generally be formed. We present the example of *Hrbaček Set Theory* (HST for short) and refer to [110], pp. 12–21.

Three classes play a key role in HST:
- $\mathbb{S} := \{x \mid \text{st}(x)\}$ (the class of standard sets),
- $\mathbb{I} := \{x \mid \text{int}(x)\}$ (the class of internal sets), where

$$\text{int}(x) \quad :\Leftrightarrow \quad \exists^{\text{st}} y \, x \in y,$$

- $\mathbb{WF} := \{x \mid \text{wf}(x)\}$ (the class of well-founded sets), where

$$\text{wf}(x) \quad :\Leftrightarrow \quad (x \neq \emptyset \Rightarrow \exists y \in x \, x \cap y = \emptyset).$$

As it turns out, all three classes interpret ZFC, i. e., the axioms of ZFC hold in all three classes (relativized to the respective class). The universe of HST is denoted by $\mathbb{H}$.

The axioms of HST are as follows:
- the axioms of extensionality, pairing, union, and infinity as in ZFC,
- the axiom schema of separation

$$\forall X \exists Y \forall x \, (x \in Y \Leftrightarrow x \in X \wedge \varphi(x)) \tag{3.27}$$

for each $\in$-st-formula $\varphi(x)$ (possibly with parameters),
- the axiom schema of collection[32]

$$\forall X \exists Y \forall x \in X \, (\exists y \, \varphi(x,y) \Rightarrow \exists y \in Y \, \varphi(x,y)) \tag{3.28}$$

for each $\in$-st-formula $\varphi(x)$ (possibly with parameters),
- φ^{st} for each axiom φ of ZFC.
- Transfer: For each $\in$-formula $\varphi(x_1, \ldots, x_n)$ with no other parameters,

$$\forall^{\text{st}} x_1 \ldots \forall^{\text{st}} x_n \, (\varphi^{\text{st}}(x_1, \ldots, x_n) \Leftrightarrow \varphi^{\text{int}}(x_1, \ldots, x_n)). \tag{3.29}$$

- Transitivity of $\mathbb{I}$:

$$\forall^{\text{int}} x \, \forall y \, (y \in x \Rightarrow \text{int}(y)). \tag{3.30}$$

- Regularity over $\mathbb{I}$: For all nonempty sets X, there is $x \in X$ with $x \cap X \subseteq \mathbb{I}$ (the full regularity of ZFC requires $x \cap X = \emptyset$ instead).
- Standardization:

$$\forall X \exists^{\text{st}} Y \, X \cap \mathbb{S} = Y \cap \mathbb{S}. \tag{3.31}$$

---

[32] The collection schema implies the replacement schema.

The set Y is uniquely determined (due to transfer and extensionality) and is denoted by ^SX.

- Saturation: The class $\mathbb{I}$ is *standard size saturated*, i. e., if $\mathcal{X} \subseteq \mathbb{I}$ is a $\cap$-closed[33] set of standard size[34] and every $X \in \mathcal{X}$ is nonempty then $\bigcap \mathcal{X} \neq \emptyset$.
- Standard size choice: If X is a set of standard size, F is a function defined on X, and $F(x) \neq \emptyset$ for any $x \in X$, then there is a function f defined on X so that $f(x) \in F(x)$ for all x (i. e., *choice* in the case when the domain of the choice function is a set of standard size).
- Dependent choice: An ω-sequence of choices exists in the case when the domain of the nth choice depends on the result of the $(n-1)$th choice (cf. Section 5.4.10).

Remarks. 1. Unlike in IST, in HST the axiom schemas of separation and collection (and thus also of replacement) apply not only to $\in$-formulas, but even to $\in$-st-formulas.
2. The axioms of power set, foundation, and choice from ZFC do not apply generally in HST (this would lead to contradictions), but only relativized to standard sets.
3. Due to the regularity over $\mathbb{I}$, the internal sets form in a certain way a basement of the HST universe. The class $\mathbb{I}$ itself is not well-founded, i. e., there are nonempty sets $X \subseteq \mathbb{I}$ without an $\in$-minimal element (e. g., the set of nonstandard $\mathbb{I}$-natural numbers).
4. We have $\mathbb{S} \subseteq \mathbb{I}$. Both $\mathbb{WF} \cap \mathbb{S}$ as well as $\mathbb{WF} \cap \mathbb{I}$ is the class of hereditarily finite sets.

Like BST, HST is also standard core interpretable. This means that $\mathbb{S}$ interprets ZFC. Because of (3.29), $\mathbb{I}$ also interprets ZFC. HST allows yet another interpretation of ZFC. We can define a $\in$-isomorphism $*$ from $\mathbb{WF}$ to $\mathbb{S}$ by $\in$-induction through $^*w := {}^S\{^*u \mid u \in w\}$. Therefore, $\mathbb{WF}$ also interprets ZFC in HST.

The following theorem summarizes the main results

Theorem 57 ([110], p. 17). *The classes $\mathbb{WF}$ and $\mathbb{S} \subseteq \mathbb{I}$ have the following properties:*
1. *The relation $\in \restriction \mathbb{S}$ is well-founded.[35] Thus $\mathbb{S}$ interprets ZFC.*
2. *The relation $\in \restriction \mathbb{WF}$ is well-founded. The class $\mathbb{WF}$ is transitive, $\subseteq$-complete[36] (moreover, $X \subseteq \mathbb{WF} \Rightarrow X \in \mathbb{WF}$), and interprets ZFC. The mapping $*$ is an $\in$-isomorphism of $\mathbb{WF}$ onto $\mathbb{S}$.*
3. *The class $\mathbb{I}$ is transitive and interprets ZFC. The mapping $*$ is an $\in$-elementary embedding of $\mathbb{WF}$ in $\mathbb{I}$, so that the following $*$-transfer holds:*

$$\forall^{\mathrm{wf}} x_1 \ldots \forall^{\mathrm{wf}} x_n \quad (\varphi^{\mathrm{wf}}(x_1, \ldots, x_n) \Leftrightarrow \varphi^{\mathrm{int}}(^*x_1, \ldots, {}^*x_n)),$$

---

[33] A set $\mathcal{X}$ is called be $\cap$-closed if $X \cap Y \in \mathcal{X}$ holds for any $X, Y \in \mathcal{X}$.
[34] *Sets of standard size* are sets of the form $\{f(x) \mid x \in X \cap \mathbb{S}\}$, where X is any set and f is any function with $X \cap \mathbb{S} \subseteq \mathrm{dom}(f)$.
[35] The restriction $\in \restriction X$ of the predicate $\in$ to a set or class X is called well-founded if every nonempty subset of X contains an $\in$-minimal element.
[36] A set or class X is called $\subseteq$-complete if every subset of X is also an element of X.

where $\varphi(x_1, \ldots, x_n)$ is an arbitrary $\in$-formula with no other parameters.

The set ω (also denoted as $\mathbb{N}_0$)[37] is defined (as in ZFC) as the smallest limit ordinal (see Section 5.2.2). Its elements are called *natural numbers*. A set X is called *finite* if there is a bijection from $n \,(:= \{0, \ldots, n-1\})$ to X for some $n \in \omega$. We call $^*\omega$ (the image of ω under the mapping $*$) the set of **-natural numbers*. A set X is called **-finite* or *hyperfinite* if there is an internal bijection from $n \,(:= \{0, \ldots, n-1\})$ to X for some $n \in {}^*\omega$ (cf. [110], p. 26). The further development of analysis is similar to the model-theoretical approach.

3.6.3 Relative set theories

In relative set theories, the predicate *standard* is not unary, but binary. A set x is therefore standard or not *relative to another set y*. This idea goes back to Péraire, who defined a *Relative Internal Set Theory* (RIST) based on Nelson's IST (see [170]). Further relative set theories come from Hrbaček [101, 102].

Relative Bounded Set Theory (RBST) by Hrbaček, which we present here, is the relative variant of BST and the basis of the textbook "Analysis with ultrasmall numbers" [105], from whose appendix we take the following explanations of RBST. The binary predicate *standard* is denoted there by the symbol $\sqsubseteq$ and satisfies the following axioms of relativization.

Axiom (relativization).

$$\forall p \, p \sqsubseteq p, \tag{3.32}$$

$$\forall p \forall q \forall r \, (p \sqsubseteq q \land q \sqsubseteq r \Rightarrow p \sqsubseteq r), \tag{3.33}$$

$$\forall p \forall q \, (p \sqsubseteq q \lor q \sqsubseteq p), \tag{3.34}$$

$$\forall p \, \emptyset \sqsubseteq p, \tag{3.35}$$

$$\forall p \exists q \, (p \sqsubseteq q \land \neg q \sqsubseteq p). \tag{3.36}$$

Thus $\sqsubseteq$ has the properties of a *total quasiorder* on the universe. It is *total* by (3.34), *reflexive* by (3.32), and *transitive* by (3.33).[38] By (3.35), $\emptyset$ is a *smallest* element and, by (3.36), there is *no largest* element.

In addition to the cumulative hierarchy (see Section 5.4.2), the universe of RBST is thus organized hierarchically in another dimension, namely in "levels of standardness" or—which sounds somewhat nicer—in levels of "observability". In [105], $p \sqsubseteq q$ is read as "p is observable relative to q". We can imagine this hierarchy in such a way that at each level the sets of the same level and those of lower levels are observable.

---

[37] In [110] it is $\mathbb{N}$, since there 0 belongs to $\mathbb{N}$.
[38] In contrast to a (partial) order, a quasiorder does not have to be *antisymmetric*, i. e., $p \sqsubseteq q \land q \sqsupseteq p$ does not generally imply $p = q$.

For the further axioms, it is convenient to write $\mathrm{st}_p(q)$ or (in the class notation) $q \in \mathrm{st}_p$ instead of $q \sqsubseteq p$. Under st_p, one can then imagine the universe up to the level of p, i.e., the class of all sets that are observable relative to p.

In RBST, the axioms of BST (i.e., ZFC plus *boundedness, transfer, standardization*, and *bounded idealization*) are adopted in a "relative version", i.e., each with the predicate st_p instead of st, and for all p. For example, the transfer schema from BST (in the version from Theorem 51) was

$$\forall^{\mathrm{st}} x_1 \ldots \forall^{\mathrm{st}} x_n \, (\varphi(x_1, \ldots, x_n)^{\mathrm{st}} \Leftrightarrow \varphi(x_1, \ldots, x_n)), \tag{3.37}$$

for each ϵ-formula $\varphi(x_1, \ldots, x_n)$. Its "relative version" then looks like this:

Axiom schema (relative transfer).

$$\forall p \, \forall^{\mathrm{st}_p} x_1 \ldots \forall^{\mathrm{st}_p} x_n \, (\varphi(x_1, \ldots, x_n)^{\mathrm{st}_p} \Leftrightarrow \varphi(x_1, \ldots, x_n)), \tag{3.38}$$

for each ϵ-formula $\varphi(x_1, \ldots, x_n)$.

In particular, we have $\forall p \, (\varphi^{\mathrm{st}_p} \Leftrightarrow \varphi)$ for each ϵ-sentence φ.[39]

We say that q is observable relative to $p_1, \ldots, p_n$, if q is observable relative to p_i for at least one $i \in \{1, \ldots, n\}$. It follows that q is observable relative to $p_1, \ldots, p_n$ if and only if q is observable relative to $(p_1, \ldots, p_n)$ (cf. [105], p. 279).

RBST differs from IST and BST in that the concepts defined with the predicate *standard* such as *ultrasmall, ultraclose, ultralarge, observable neighbor*,[40] etc. are defined *relative to a context $p_1, \ldots, p_n$*. They are therefore called *relative concepts*.

Definition 20. Let $p := (p_1, \ldots, p_n)$. Then
1. $x \in \mathbb{R}$ is called *ultrasmall* relative to $p_1, \ldots, p_n$ if $x \neq 0$ and

$$\forall^{\mathrm{st}_p} y \, (y \in \mathbb{R} \wedge y > 0 \Rightarrow |x| < y).$$

2. $x, y \in \mathbb{R}$ are said to be *ultraclose* ($x \approx y$) relative to $p_1, \ldots, p_n$ if $x - y$ is 0 or ultrasmall relative to $p_1, \ldots, p_n$.
3. $x \in \mathbb{R}$ is called *ultralarge* relative to $p_1, \ldots, p_n$ if

$$\forall^{\mathrm{st}_p} y \, (y \in \mathbb{R} \Rightarrow |x| > y).$$

Theorem 58. *For all $x \in \mathbb{R}$, if x is not ultralarge relative to $p_1, \ldots, p_n$, then there is a unique $y \in \mathbb{R}$ with $\mathrm{st}_{(p_1, \ldots, p_n)}(y)$ and $x \approx y$ relative to $p_1, \ldots, p_n$. This y is called the observable neighbor of x relative to $p_1, \ldots, p_n$.*

---

[39] Thus, it is again irrelevant whether the ZFC axioms or the ZFC axioms relativized by st_p (for all p) are postulated in RBST (cf. footnote 30).

[40] In [105] these terms are used instead of *infinitesimal, infinitely close, unlimited, standard part*. The authors also write $\simeq$ instead of $\approx$. In contrast to our use of *infinitesimal, ultrasmall* excludes 0.

The proof of Theorem 58 is analogous to the proof of Theorem 56 in IST (or BST), except that the steps are always carried out relative to the context. The rules for calculating with $\approx$ and standard parts (observable neighbors) apply accordingly (i. e., always relative to the context).

In IST, for example, we have the following rule: For all $x, y, a, b \in \mathbb{R}$, if x, y are limited and $x \approx a$ and $y \approx b$, then $xy \approx ab$.

In RBST this becomes the following rule:

Proposition. *For all $x, y, a, b \in \mathbb{R}$,*

$$\begin{aligned} &\textit{if} \quad \textit{not } x, y \textit{ ultralarge relative to } p_1, \ldots, p_n \quad \textit{and} \\ &\qquad x \approx a \textit{ relative to } p_1, \ldots, p_n \quad \textit{and} \\ &\qquad y \approx b \textit{ relative to } p_1, \ldots, p_n, \\ &\textit{then} \quad xy \approx ab \textit{ relative to } p_1, \ldots, p_n, \end{aligned} \quad (3.39)$$

where $p_1, \ldots, p_n$ is any context.

The explicit specification of a context is very cumbersome in the long run and initially seems to be a major disadvantage of RBST. In practice, however, this disadvantage is circumvented by the following convention.

Convention about contexts: In a theorem, definition, or proof, whenever a relative concept is used without explicit specification of its context, it is understood to be relative to the context of that theorem, definition, or proof ([105], p. 27).

We illustrate this using the example of the definition of continuity.

Definition 21. Let $f : D \to \mathbb{R}$ be a function and $a \in D$. Then f is said to be *continuous* at a if for all $x \in D$,

$$x \approx a \Rightarrow f(x) \approx f(a). \quad (3.40)$$

At first glance, this definition looks like the definition of S-continuity in Section 2.10.2. However, due to the convention about contexts, (3.40) is to be understood *relative to the context of the definition*. In this case, the context consists of the parameters f and a. Therefore, (3.40) means in detail:

$$x \approx a \text{ relative to } f, a \quad \Rightarrow \quad f(x) \approx f(a) \text{ relative to } f, a. \quad (3.41)$$

The advantage of RBST over IST or BST is that Definition 21 is directly equivalent to ε–δ-continuity. With S-continuity, this was only the case for standard parameters, so that a general definition of continuity was only obtained from S-continuity through standardization (see Section 3.4.9).

For example, the function $f(x) = x^2$ (which was a counterexample for S-continuity at unlimited points) is continuous at *all* points a in the sense of Definition 21 because

$x \approx a$ (relative to a) implies that x is not ultralarge (relative to a) and hence, by (3.39), $x^2 \approx a^2$ (relative to a).

In RBST, an $\{\in, \sqsubseteq\}$-formula $\varphi(x_1, \ldots, x_n)$ is called *internal* if all occurrences of the predicate *standard* are relative to the parameters $x_1, \ldots, x_n$, in other words, if it arises from an $\{\in, \text{st}\}$-formula by replacing all occurrences of st with $\text{st}_{(x_1,\ldots,x_n)}$. A concept is called *internal* if it is defined by an internal formula. Like continuity, the derivative, integral and limit are defined by internal formulas and are therefore internal concepts. When operating with internal concepts, the context no longer appears explicitly (due to the convention about contexts). The further development of elementary analysis based on RBST is explained in [105] (see also Section 4.2.6).

4 On practice and acceptance in teaching

4.1 Textbooks on analysis

The textbooks that are usually recommended as literature in calculus or analysis lectures (e. g., [5, 18, 128, 150, 187, 188, 209]) all take a classical approach based on Weierstrass's notion of a limit. The real numbers are introduced axiomatically as a complete ordered field, questions about existence and uniqueness, if addressed, are usually answered with references to the literature. In [188] the construction of the real numbers (starting from the rational numbers) is carried out in the appendix of Chapter 1 (without showing uniqueness). Spivak shows the construction and the uniqueness of the real numbers in an epilogue ([209], pp. 578–598).

There are not many textbooks that take a nonstandard approach to calculus. The first textbook of this kind was *Elementary Calculus* by Keisler. The first edition was published in 1976 [118], the second in 1986. A revised version from 2012 is freely available online [120], and in print as a third edition [119]. With almost 1000 pages, this book is very extensive and detailed. Much slimmer (135 pages) is the *Infinitesimal calculus* by Henle and Kleinberg [90]. Both textbooks introduce hyperreal numbers axiomatically (see Section 4.2.4).

The omega calculus presented in Section 3.2 has influenced Henle's "Non-nonstandard Analysis" [89] and Tao's online post "A cheap version of nonstandard analysis" [219]. Both contributions show briefly and exemplarily what is possible with a construction based on the Fréchet filter Cof in infinitesimal calculus. However, they are not complete textbooks. Another non-Archimedean extension of $\mathbb{R}$, which can be constructed without ultrafilters is the system of *superreal numbers* by David O. Tall (see [217]).

An introductory calculus textbook (in French) based on Nelson's internal set theory (see Section 3.5) is [60] by Deledicq and Diener. It is only available in antiquarian bookshops. Robert's book [178] is a concise and readable introduction to internal set theory and deals with elementary analysis in the first part. Overall, it is aimed more at advanced students. The textbook [105] by Hrbaček, Lessmann, and O'Donovan uses a simplified version of *Relative Bounded Set Theory* RBST (see Section 3.6.3) and is intended for teaching calculus at a high school or college level.

4.2 Nonstandard introductions to analysis

4.2.1 Construction with Fréchet filter

In Section 3.2 we presented a genetic construction of analysis with infinitesimals according to [129]. The basis there was the set $^{\Omega}\mathbb{R} := \mathbb{R}^{\mathbb{N}}/\text{Cof}$ of omega numbers constructed with the Fréchet filter Cof.

The advantages of the omega calculus lie in the simple construction of the omega numbers and the simple extension of all real functions and relations. Furthermore, equations and inequalities, as well as their conjunctions, can be transferred from $\mathbb{R}$ to $^\Omega\mathbb{R}$, which is already sufficient for many considerations. The main disadvantage is that $^\Omega\mathbb{R}$ is not an ordered field, but only a partially ordered ring with nontrivial zero divisors. Only a limited transfer principle applies.

As far as the question of the suitability of omega numbers for an introduction to calculus is concerned, the disadvantages described are less disturbing when calculating than when visualizing. In particular, due to the lack of total order, the omega numbers cannot be visualized linearly ordered on a straight line. At the "formula level", however, these numbers offer advantages, as definitions and proofs are simpler compared to standard analysis.

In [89], Henle simplifies even further compared to [129] by dispensing with the formation of equivalence classes (which may already be quite abstract for beginners) and working directly with the sequences. In this way, the direct reference to standard analysis is retained, but at the same time the suggestive notation of nonstandard analysis can be used.

Henle uses bold letters to distinguish sequences from real numbers, for example, $\mathbf{a} := (a_n)_{n \in \mathbb{N}}$. It is permitted that the sequence is not defined for a finite number of indices. Equations or inequalities with sequences (for example, $\mathbf{a} = \mathbf{b}$, $\mathbf{a} = 0$, $\mathbf{a} \neq 0$, $\mathbf{a} > 0$), as well as membership relations (for example, $\mathbf{a} \in D$ with $D \subseteq \mathbb{R}$) are to be interpreted in such a way that they hold "from some point on", i.e., for all $n \geq k$ for some k (or "almost everywhere" as we said in Section 3.2). This stipulation turns the "equality" of sequences into an *equivalence relation*, but Henle refrains from explicitly defining it this way. Care should be taken when using the symbol $\neq$. For example, $\mathbf{a} \neq 0$ *does not* mean "not $\mathbf{a} = 0$" but "$a_n \neq 0$ from some point on". Applying a function f to a sequence $\mathbf{a}$ means applying it to the terms of the sequence, i.e., $f(\mathbf{a}) := (f(a_n))_{n \in \mathbb{N}}$ for all $\mathbf{a} \in \text{dom}(f)$.

Analogous to nonstandard analysis with hyperreal numbers, one defines that a sequence $\mathbf{a}$ is
- *infinitely small* ($\mathbf{a} \approx 0$) if $|\mathbf{a}| < r$ holds for all positive $r \in \mathbb{R}$,
- *finite* or *bounded* if there is $r \in \mathbb{R}$ such that $|\mathbf{a}| < r$.

Also the following concepts for functions $f : D \to \mathbb{R}$, $D \subseteq \mathbb{R}$ are defined analogously to nonstandard analysis with hyperreal numbers:
- f is *continuous* in $r \in D$ if for all $\mathbf{a} \in D$,

$$\mathbf{a} \approx r \Rightarrow f(\mathbf{a}) \approx f(r).$$

- f is *uniformly continuous* on D if for all $\mathbf{a}, \mathbf{b} \in D$,

$$\mathbf{a} \approx \mathbf{b} \Rightarrow f(\mathbf{a}) \approx f(\mathbf{b}).$$

- For $r, d \in \mathbb{R}$, we say $f'(r) = d$ if there is a $\mathbf{\Delta x} \neq 0$ with $\mathbf{\Delta x} \approx 0$ and $r + \mathbf{\Delta x} \in D$ and if for all such $\mathbf{\Delta x}$ we have

$$\frac{f(r + \mathbf{\Delta x}) - f(r)}{\mathbf{\Delta x}} \approx d.$$

The definitions have become so simple compared to the standard definitions because the otherwise necessary quantifiers are contained in the definition of the relation "$\approx$". In fact, $\mathbf{a} \approx 0$ means by definition nothing other than

$$\forall \varepsilon > 0 \, \exists n_0 \in \mathbb{N} \, \forall n \in \mathbb{N} \, (n \geq n_0 \Rightarrow |a_n| < \varepsilon).$$

The lack of total order has the effect that sometimes subsequences must be considered. Henle writes $\mathbf{a} \subset \mathbf{b}$ for "$\mathbf{a}$ is a subsequence of $\mathbf{b}$". Instead of the standard part principle (valid for the hyperreals), we have the following

Proposition. *If $\mathbf{a}$ is finite, then for some $\mathbf{c} \subset \mathbf{a}$ and some $r \in \mathbb{R}$,*

$$\mathbf{c} \approx r.$$

This corresponds to the Bolzano–Weierstrass theorem in standard analysis. The proof depends on the completeness of $\mathbb{R}$. Henle uses the identification of the real numbers with the infinite decimal fractions for this purpose.

The need to operate with subsequences exists, for example, when proving the chain rule, the intermediate value theorem or the statement that every continuous function is integrable (see [89]).

Henle uses sequences of step functions to define the integral.

Definition 22. For a function f and an interval $[p, q]$,

$$\int_p^q f \, dx = r$$

if there are sequences $\mathbf{d}, \mathbf{u}$ of step functions on $[p, q]$ with $\mathbf{d} \leq f \leq \mathbf{u}$ and

$$\int_p^q \mathbf{d} \, dx \approx r \approx \int_p^q \mathbf{u} \, dx.$$

Step functions and the integral of step functions are defined as usual. The integral of a sequence of step functions is defined as the sequence of integrals of the step functions. The inequality $\mathbf{d} \leq f \leq \mathbf{u}$ means $d_n \leq f \leq u_n$ from some point on.

4.2.2 Generalized Ω-adjunction

The disadvantages of the construction $\mathbb{R}^\mathbb{N}/\text{Cof}$ as an extension of $\mathbb{R}$ described above are overcome if a free ultrafilter $\mathcal{U}$ is used instead of Cof (see Section 3.3.1). The resulting extension $\mathbb{R}^\mathbb{N}/\mathcal{U}$ is an ordered field and we have the full transfer principle for first-order sentences (see Theorem 45). However, a construction using ultrafilters does not seem very suitable for beginners' courses.

In [130], Laugwitz offers an alternative approach which, although not completely constructive, avoids explicit recourse to ultrafilters. This way consists of a generalized adjunction of an infinitely large element Ω and the postulation of a principle with which true statements about the extended number system can be obtained. Laugwitz calls it the *Leibniz's principle* in reference to a formulation from a letter from Leibniz to Varignon, according to which the rules of the finite also succeed in the infinite (see the Leibniz quote on p. 15). The procedure is formulated for an arbitrary Archimedean field K.

Adjunction of Ω. Each sequence $a(n) \in K$, defined for all sufficiently large natural n, or, in other words, for all $n \geq n_0$ with an $n_0 \in \mathbb{N}$, specifies an element of the extended number system $^\Omega K$; we write $a(\Omega)$ for this element and call it an omega number ([130], p. 85).

Leibniz's principle. Let $A(\cdot)$ be a statement formulated in the language of K. If there is an $n_0 \in \mathbb{N}$ such that for all $n \geq n_0$ the statement $A(n)$ is true in the underlying theory of K, then $A(\Omega)$ should be included as a true statement in the new theory of $^\Omega K$ ([130], p. 88).

The adjunction of Ω given here is a generalization of the field extension $K(\Omega)$, since $a(n)$ can be any sequence and not just a rational fraction in n. This means, for example, that $(1 + \frac{1}{\Omega})^\Omega$ or $\sum_{k=0}^\Omega \frac{1}{k!}$ are also omega numbers. The "language of K" to which Leibniz's principle refers is the first-order language with constants for each element of K and function and relation symbols for all functions and relations on K.[1]

Leibniz's principle proves that $^\Omega K$ is an ordered field. The existence of the multiplicative inverse and the trichotomy of the order relation $<$ are particularly interesting here because these axioms do not hold in $K^\mathbb{N}/\text{Cof}$. If a sequence $a(n) \in K$ is given, then for sufficiently large (even for all) n,

$$a(n) = 0 \vee \exists x\, a(n) \cdot x = 1.$$

Hence, according to Leibniz's principle,

$$a(\Omega) = 0 \vee \exists x\, a(\Omega) \cdot x = 1.$$

---

[1] Laugwitz also allows symbols from set theory (for example, $\in$, $\cup$, $\cap$). Since only individual variables (for elements of K) are available, this is not a substantial extension of the language. However, one can, for example, write the familiar $x \in \mathbb{N}$ instead of the unfamiliar $\mathbb{N}x$.

If sequences $a(n)$ and $b(n)$ are given, then for all sufficiently large (even for all) n,

$$a(n) = b(n) \vee a(n) < b(n) \vee a(n) > b(n).$$

Hence, according to Leibniz's principle,

$$a(\Omega) = b(\Omega) \vee a(\Omega) < b(\Omega) \vee a(\Omega) > b(\Omega).$$

However, it is not always possible to decide which of these options applies for specific omega numbers.

Using the example of $(-1)^\Omega$, the application of Leibniz's principle means: Since $(-1)^n = 1 \vee (-1)^n = -1$ holds in K for sufficiently large (even for all) n, $(-1)^\Omega = 1 \vee (-1)^\Omega = -1$ holds in $^\Omega K$. Which of the possibilities applies remains undetermined.

The indeterminacy of Ω in the theory of $^\Omega K$ does not interfere with the further construction of the analysis, just as the indeterminacy of the ultrafilter $\mathcal{U}$ in the construction $^*K := K^\mathbb{N}/\mathcal{U}$ does not interfere with the construction of the analysis with $^*K$. For example, it is not important whether $\mathcal{U}$ contains the set of even or the set of odd numbers. Laugwitz notes:

> The theory of $^\Omega K$ is, so to speak, the common core of all possible nonstandard theories of the $^*K$. It contains all the infinitesimal mathematics which is independent of the arbitrary choice of a special ultrafilter and even of the existence of the ultrafilters themselves ([130], p. 103).[2]

4.2.3 Superreal numbers

The simplest way to extend $\mathbb{R}$ to a non-Archimedean field is to adjoin a new element that is by definition infinitely large (larger than any real number) or alternatively infinitely small (positive but smaller than any positive real number). This approach is generalized with the superreal numbers by including certain series in ε in addition to a new, infinitely small element ε (cf. [217]). Tall defines his superreal numbers as formal Laurent series in ε, i. e., as series of the form

$$\sum_{i=m}^{\infty} a_i \varepsilon^i \tag{4.1}$$

with $m \in \mathbb{Z}$ and $a_i \in \mathbb{R}$. "Formal" means that one is not concerned with convergence. Technically, the superreal numbers can be identified with those elements of $\mathbb{R}^\mathbb{Z}$ for

---

2 In the original German: "Die Theorie von $^\Omega K$ ist sozusagen der gemeinsame Kern aller möglichen Nichtstandard-Theorien zu den $^*K$. Sie enthält alle diejenige Infinitesimalmathematik, welche von der willkürlichen Wahl eines speziellen Ultrafilters und sogar von der Existenz der Ultrafilter selbst unabhängig ist."

which only finitely many terms with a negative index are nonzero. The representation as a series in ε is used for intuition, for example, to motivate the definitions of addition, multiplication, and order of superreal numbers. Here, ε is to be thought of as an infinitely small number.

Tall denotes the set of all superreal numbers by $\mathfrak{R}$. Then $\mathbb{R}$ is a subset of $\mathfrak{R}$ and the real numbers in $\mathfrak{R}$ are exactly those for which all coefficients except a_0 vanish in (4.1). Addition, multiplication, ordering, and the additive inverse are defined in an obvious way (by formally operating with the series of type (4.1)). The multiplicative inverse for $\alpha \in \mathfrak{R} \setminus \{0\}$ exists. If $\alpha = \sum_{i=m}^{\infty} a_i \varepsilon^i$ with $a_m \neq 0$ is given, then there is $\beta = \sum_{j=-m}^{\infty} b_j \varepsilon^j$ with $\alpha\beta = 1$ because the coefficients b_j can be determined recursively from

$$\left(\sum_{i=m}^{\infty} a_i \varepsilon^i\right) \cdot \left(\sum_{j=-m}^{\infty} b_j \varepsilon^j\right) = \sum_{k=0}^{\infty} \sum_{j=0}^{k} a_{m+j} b_{-m+k-j} \varepsilon^k = 1.$$

We have $b_{-m} = a_m^{-1}$ and for $k \geq 1$,

$$b_{-m+k} = -a_m^{-1} \sum_{j=1}^{k} a_{m+j} b_{-m+k-j}.$$

The set $\mathfrak{R}$ thus becomes an ordered field. Since $\varepsilon^{-1} > n$ holds for all $n \in \mathbb{N}$, $\mathfrak{R}$ is a non-Archimedean extension field of $\mathbb{R}$.

Finite, infinite, and infinitesimal numbers, as well as the standard part of finite numbers, are defined as usual. The finite superreal numbers are precisely those for which all coefficients with a negative index vanish. The standard part of a finite superreal number is the coefficient with index zero.

Every analytic function $f : D \to \mathbb{R}$ (D open interval) can be extended to a function $f : D^\# \to \mathfrak{R}$ with $D^\# = \{x \in \mathfrak{R} \mid \mathrm{stp}(x) \in D\}$ by defining for each $\delta \approx 0$,

$$f(x + \delta) := \sum_{n=0}^{\infty} a_n \delta^n,$$

where $f(x + h) = \sum_{n=0}^{\infty} a_n h^n$ is the power series expansion of f around the center $x \in D$ (and h is sufficiently small). By inserting the series representation of δ and multiplying it out, one obtains the series representation of $f(x + \delta)$ as a series in ε.

For analytical functions, the terms *continuity* and *derivative* can be defined as usual in nonstandard analysis, Then, for example, the differentiation rules can be deduced and the derivatives of specific functions can be calculated. Similar to Keisler, definitions or proof situations can be illustrated geometrically using "microscopes" or "telescopes" by choosing a suitable magnification or reduction factor (the appropriate ε power).

Of course, there are some limitations compared to basic analysis with hyperreal numbers:
- Only analytic functions can be considered.

- There are no infinitely large integers available.
- There is no general transfer principle for first-order statements.

Due to the lack of infinitely large integers, sequences and series cannot be generalized in the way that is possible in the hyperreals (with hyperintegers as indices and summation limits). Integrals cannot be defined using hyperfinite Riemann sums.

Tall defines the integral of analytical functions using the concept of the *area function* (see Definition 23) and thus proves the fundamental theorem.

Definition 23 ([217], p. 43). We define an *area function* A_f for an analytic function $f : D \to \mathbb{R}$ (where D is an open interval) to be a superreal function $A_f(u, v)$ of two variables u, v in $D^\#$ such that
1. $A_f(u, v) + A_f(v, w) = A_f(u, w)$ where $u, v, w \in D^\#$,
2. if $x \in D^\#$ and θ is a nonzero infinitesimal, then $\frac{A_f(x, x+\theta)}{\theta} \approx f(x)$.

The fundamental theorem is formulated with area functions as follows (cf. [217], p. 44):
- If $f : D \to \mathbb{R}$ is analytic on the open interval D, A_f is an area function for f, and for some $a \in D$ we define $F(x) := A_f(a, x)$ (for $x \in D^\#$), then $F' = f$.
- Conversely, if the analytic function F satisfies $F' = f$, then $A_f(a, b) := F(b) - F(a)$ is an area function.

For the proof, see [217], p. 44.

4.2.4 Axiomatic introduction of hyperreal numbers

Keiser's *Elementary Calculus* is "aimed at the average beginner calculus student and covers the usual three or four semester sequence" ([119], p. iv, preface to the first edition). It is based on the following three principles:

Extension principle. 1. The real numbers form a subset of the hyperreal numbers, and the order relation $x < y$ for the real numbers is a subset of the order relation for the hyperreal numbers.
2. There is a hyperreal number that is greater than zero but less than every positive real number.
3. For every real function of one or more variables, we are given a corresponding hyperreal function $^*f$ of the same number of variables. This $^*f$ is called the *natural extension* of f.

([119], pp. 27–28.)

Transfer principle. Every real statement that holds for one or more particular real functions holds for the hyperreal natural extensions of these functions ([119], p. 28).

Keisler defines a real statement as a combination of equations or inequalities of real formulas, as well as statements that specify whether a real formula is defined or not.

Standard part principle. Every finite hyperreal number is infinitely close to exactly one real number ([119], p. 36).

The finite hyperreal numbers are those that are between two real numbers (see [119], p. 30). The unique real number that is infinitely close to the finite hyperreal number b is called the *standard part* of b.

A more formal version of the above principles can be found in [121] with the axioms A–E. There, a construction of hyperreal numbers using ultrafilters is also carried out (Chapter 1G) and the proof is provided that the restriction to combinations of equations and inequalities in the transfer principle is not essential and that the (apparently more general) elementary extension principle for first-order sentences (see Section 4.4.5) follows from the axioms A–E (Chapter 15A).

In *Elementary Calculus*, Keisler introduces the hyperintegers as the function values of the natural extension of the *greatest integer function* $x \mapsto [x]$, where $[x]$ denotes the greatest integer n such that $n \leq x$ ([119], p. 159). More generally, the extension principle formulated for functions can be used to extend any relation $P \subseteq \mathbb{R}^n$ by considering its characteristic function $1_P : \mathbb{R}^n \to \{0, 1\}$ (with $1_P(x) = 1 \Leftrightarrow x \in P$) and defining

$$^*P := \{x \in {}^*\mathbb{R}^n \mid {}^*1_P(x) = 1\}$$

(cf. [121], p. 14).

In their *Infinitesimal Calculus*, Henle and Kleinberg give a brief (not too formal) introduction to languages and structures (partly with examples outside mathematics) and then define a language L for describing the structure of the real numbers (also called *the system of real numbers*). In addition to the usual logical symbols, L contains constants for all real numbers, as well as function and relation symbols for all functions and relations on the real numbers. It is therefore clear that the *alphabet* of this language cannot be explicitly specified, but is an abstract set.

A structure S is a *hyperreal number system*, if it has the following three properties (cf. [90], p. 25):[3]

1. *S contains the real number system.* This does not only mean that all real numbers are in S, but also that every function and relation defined on reals is also defined on numbers in S.
2. *S contains an infinitesimal.* That is, there is a number in S that is greater than 0 but smaller than every positive real number.
3. *The same sentences of L are true in both S and $\mathbb{R}$.* If B is any sentence of L, then B is true in S if and only if B is true in $\mathbb{R}$.

---

[3] Henle and Kleinberg point out that S is not uniquely determined by these properties and that the indefinite article in "a hyperreal number system" is therefore appropriate.

Henle and Kleinberg do not stop at the mere definition of the term *hyperreal number system*, but also explain how a specific such system (which they denote by ℍR) can be constructed, although the more demanding parts (the proof of existence of ultrafilters and the proof of the transfer principle) are relegated to the appendix. They leave it up to the reader to skip the construction of ℍR since knowledge of the construction is not needed for the rest of the book. A similar situation exists in standard analysis when the real numbers are introduced axiomatically and the possibility of their construction is only mentioned in textbooks or the construction is carried out in the appendix (as in [188]).

The definitions of continuity, derivative and limit are analogous to the definitions given in Definition 10 in Section 3.2.5. Only the definition of the integral differs, since the concept of hyperfinite sums is missing. Instead of general Riemann sums, sums with equidistant interval partitions are used and the sum (as a function of the step size) is extended to hyperreal numbers so that it is also defined for infinitesimal step sizes.

To be more precise, this is done as follows: For a real function $f : [a, b] \to \mathbb{R}$ and real $h > 0$, let

$$S_a^b(f, h) := \sum_{i=0}^{n-1} f(x_i)h + f(x_n)(b - x_n),$$

where $n := \max\{i \in \mathbb{N}_0 \mid a + ih \leq b\}$, $x_i := a + ih$, for $i = 0, \ldots, n$.

If f, a, b are fixed, then $S(h) := S_a^b(f, h)$ is a real function defined on $\mathbb{R}^+$ with a hyperreal extension defined on $^*\mathbb{R}^+$. In the infinitesimal case, dx is often used instead of h.

Definition 24. Let $a, b \in \mathbb{R}$, $a < b$ and $f : [a, b] \to \mathbb{R}$. Then f is called integrable from a to b if there exists $c \in \mathbb{R}$ such that for every infinitesimal $dx > 0$,

$$S_a^b(f, dx) \approx c.$$

This c is called the *integral of f from a to b* and is denoted by $\int_a^b f(x)\,dx$.

In the case of Riemann-integrable functions, Definition 24 and the classical definition using Riemann sums lead to the same integral, but there are functions that are integrable according to Definition 24 but not Riemann-integrable (see [90], pp. 118–119). In elementary analysis (for example, when proving common integral theorems or the fundamental theorem), this fact does not cause any problems.

4.2.5 Elementary calculus based on internal set theory

The textbook [60] by Deledicq and Diener begins with a paradox related with the existence of "infinitely large" or "unlimited" natural numbers: if we assume that the set E of all limited numbers of $\mathbb{N}_0$ contains 0 and with every number n also its successor $n + 1$, then according to the principle of mathematical induction $E = \mathbb{N}_0$. Hence, all numbers in $\mathbb{N}_0$ are limited. The existence of unlimited numbers in $\mathbb{N}_0$ would therefore be a paradox.

In internal set theory, the paradox is resolved by assuming that there is no set E that contains exactly all limited numbers, that is, that the limited numbers in $\mathbb{N}_0$ cannot be collected into a set in the sense of classical mathematics.

Deledicq and Diener thus motivate the introduction of a new predicate *standard* into the language of mathematics and define *limited, unlimited* and *infinitesimal* numbers as in Section 2.6.2 (Definition 4). To make it easier to get started with internal set theory, the authors set out provisional principles for dealing with the new predicate and only present the complete system of IST axioms in the second part of the book. The provisional principles are as follows:

First principle. If M is an object that is defined without (direct or indirect) use of the predicate *standard*, possibly using other standard objects, then M is standard.

Second principle. Let M be a set defined without (direct or indirect) use of the predicate *standard*. Then all elements of M are standard if and only if M is finite.

Transfer principle. Let $P(x)$ be a property that is formulated without (direct or indirect) use of the predicate *standard*. Then $P(x)$ is true for all x if and only if $P(x)$ is true for all standard x.

The second principle corresponds to Theorem 53 from Section 3.5.3, which is proved in IST using the idealization axiom. It follows from this principle that every infinite set contains nonstandard elements and that in $\mathbb{N}_0$ all standard numbers come *before* all nonstandard numbers. The transfer principle corresponds to the transfer axiom in IST (but without parameters). If standard parameters were allowed in $P(x)$, the first principle would be a consequence of the transfer principle.

Based on the three principles mentioned above, the first part of the book introduces the concepts of S-continuity, S-limit, S-derivative, and S-integral, and proves the essential theorems. Strictly speaking, Deledicq and Diener assume a further principle, namely the

Standard part principle. Every real number is infinitely close to a standard real number.

The authors formulate this principle as a theorem, but postpone the proof to the second part, where the standardization axiom necessary for the proof is available. There, the connection between the S-concepts and the corresponding concepts of classical analysis is also discussed and the fundamental theorem is proved.

4.2.6 Elementary calculus based on relative set theory

The textbook [105] is based on *Relative Bounded Set Theory* (RBST) (see Section 3.6.3) using the so-called *standard picture* (see Section 5.4.8). This means that we imagine that the familiar universe of traditional mathematics (which, for example, contains the sets ℕ and ℝ, as well as functions and relations on these sets) is enriched with new, ideal objects (such as unlimited natural numbers and infinitesimal nonzero real numbers) within the framework of an extended mathematics. Even familiar sets, such as ℝ or ℕ, therefore contain new, ideal elements in this perspective. The truth or falsehood of statements of traditional mathematics remains unaffected. True statements remain true and false statements remain false. For example, in the extended mathematics, ℝ is also a complete ordered field.

Hrbaček, Lessmann, and Donovan use an analogy from zoology. There is a class of mammals, which includes lions, horses, bats, whales, and kangaroos, for example. In an "extended zoology", the class of mammals also includes fictional creatures (ideal elements) such as unicorns and yetis that have the same characteristics as real mammals (they are homeothermic, nurse their young, etc.) (see [105], p. 4).

In order to be able to talk about the new objects, the language of traditional mathematics is extended by a new term that has no meaning in traditional mathematics, i. e., that cannot be defined using already known terms, but remains undefined. This new term is named *observable* by the authors of the textbook, more precisely, *observable relative to*, because it is a binary relation. For all objects p, q, either "q is observable relative to p" or not. Objects that are observable relative to *any* object are called *standard* objects. Intuitively (according to the standard picture), these are all the objects of traditional mathematics, while the new, ideal objects (for example, nonzero infinitesimals) are characterized by the fact that they are *not* standard, that is, *not* observable relative to any object.

We say that q is observable relative to $p_1, \ldots, p_k$ if p is observable relative to at least one of the p_i, $i = 1, \ldots k$. We call $p_1, \ldots, p_k$ a *context*. It is therefore clear that if q is observable relative to a particular context, then it is observable relative to any extended context.

Concepts that are derived from the term *observable* and also depend on the context are called *relative concepts*. Examples are the terms *ultrasmall*, *ultralarge*, and *ultraclose*, which are used by the authors instead of the otherwise usual terms *infinitesimal*, *unlimited* and *infinitely close* (see Definition 20).

The *Convention about Contexts* already introduced in Section 3.6.3 then simplifies formulations: If relative concepts are used in theorems, definitions or proofs without explicitly stating a context, they are to be understood relative to the context of the theorem, definition or proof.

A statement is called *internal* if the context of all relative concepts occurring in it is given by the parameters of the statement.

The analysis with ultrasmall numbers is based on the following principles (see [105], pp. 32–33), which are essentially an informal version of the axioms from RBST (partly reduced to special cases) or follow from the axioms.

Relative observability principle. For all p, q and r,
1. p is observable relative to p.

2. If p is observable relative to q and q is observable relative to r, then p is observable relative to r.
3. If p is not observable relative to q, then q is observable relative to p.

Stability principle. An internal statement is equivalent to the statement obtained from it by extending its context by additional parameters.

Existence principle. There exist ultrasmall numbers.

Closure principle. Given an internal statement with parameters $p, p_1, \ldots, p_k$, if $p_1, \ldots, p_k$ are observable and there exists some object p for which the statement is true, then there exists some observable object p for which the statement is true.

Observable neighbor principle. If a real number is not ultralarge, then there is an observable real number that is ultraclose to it.

Definition principle. Internal defining statements can be used to define sets and functions. These sets and functions are observable whenever all the parameters of their definition are observable.

On the basis of these principles, the analysis is developed, as indicated in Section 3.6.3. The derivation of the principles from the axioms of RBST is given in the appendix of [105]. The last four principles are relative to a given context.

4.2.7 Summary and comparison

Table 4.1 summarizes the elementary approaches to an infinitesimal enriched analysis discussed in this chapter, along with their main advantages and disadvantages. The criterion of whether *internal sets* and, based on these, *hyperfinite sets* can be considered is not taken into account.

While this criterion is absolutely crucial for advanced applications of nonstandard analysis (which is why, for example, superreal numbers are ruled out for these purposes), elementary calculus can largely do without these more sophisticated concepts. In the axiomatic approaches of Keisler or Henle/Kleinberg, they are not readily available, and they are also completely dispensed with in the constructive approaches of Henle or Tao. One consequence of this restriction is that the integral cannot be defined using hyperfinite Riemann sums (see Definition 10 in Section 3.2.5), but rather using equidistant infinitely fine partitions (see Definition 24 in Section 4.2.4).

It is also possible to treat internal and hyperfinite sets in constructions with the Fréchet filter and in the generalized Ω-adjunction (see [129] and [130]). However, this makes the program significantly more demanding, and the question arises as to whether this effort is justified if the goal is only an infinitesimal approach to elementary calculus.

In the approaches based on internal set theories (e. g., IST, RBST), the definition of internal and hyperfinite sets is not an issue, as all sets are internal and *finite* sets can

Table 4.1: Summary and comparison of elementary nonstandard introductions to calculus.

Approach	References	Advantages	Disadvantages
Construction with Fréchet filter	[129, 89, 219]	Simple construction; direct reference to standard analysis	No ordered field; limited transfer principle
Superreal numbers	[217]	Simple construction	Restriction to analytical functions; no infinitely large "superintegers"; no transfer principle
Generalized Ω-adjunction	[130]	No construction required; transfer principle for first-order statements ("Leibniz's principle")	"Leibniz's principle" must be accepted as an axiom.
Hyperreal numbers axiomatically	[119, 90]	No construction required; transfer principle for first-order statements	Additional principles (extension, transfer, standard part) must be accepted as axioms.
Internal set theory	[60, 178]	No construction required; transfer axiom for internal statements; awareness of mathematical foundations is promoted.	Greater reference to set theory is necessary. One must beware of "illegal set formation" and "illegal transfer". Additional axioms must be accepted. Nonstandard definitions require standardization.
Relative set theory	[105]	Direct nonstandard definitions using "relative concepts". Otherwise, advantages as with internal set theory.	Extension is more complex due to the concept of "relative observability". Otherwise, disadvantages as with internal set theory (except the last point).

have an unlimited number of elements. Instead, care must be taken here not to form any "illegal sets" (see Section 3.5.1). In [105], the integral is defined using Riemann sums with arbitrary infinitely fine partitions. The authors of [60] and [178] are content with an integral definition using equidistant partitions.

4.3 Teaching experience

4.3.1 The cognitive advantage hypothesis

The first studies on the applicability of nonstandard analysis in introductory courses were conducted by Kathleen Sullivan. For her dissertation, she accompanied an experiment at four small private colleges and a larger public high school in the Chicago and Milwaukee areas and published excerpts of the results in the *American Mathematical Monthly* [210]. A control group was taught traditional calculus in 1972/73 and an experimental group was taught Keisler's *Elementary Calculus* (an earlier version from 1971) by the same teachers in 1973/74. Both groups were the same size (68 students each) and comparable in terms of prerequisites.[4]

The study included a 50-minute test in both groups, interviews with the instructors of both groups, and a questionnaire filled out by all those who had used Keisler's book within the past five years. The test examined the ability to define basic concepts, compute limits, produce proofs, and apply basic concepts. Table 4.2 shows how many students in the two groups attempted to solve the various tasks. The results were the same in both groups for the task of applying basic concepts, while for all other tasks the experimental group was ahead of the control group.

Table 4.2: Number of Students Attempting a Solution ([210], p. 373).

	Control Group (68 students)	Experimental Group (68 students)
Defining basic concepts	48	52
Computing limits	49	68
Producing proofs	18	45
Applying basic concepts	60	60

A more detailed analysis of the solutions and attempts at solutions is given in [210] only for the third task, where the difference between the two groups was particularly clear. The task was:

Define $f(x)$ by $f(x) = x^2$ for $x \neq 2$, $f(x) = 0$ for $x = 2$. Prove, using the definition of a limit, that $\lim_{x \to 2} f(x) = 4$.[5]

In the experimental group, 25 of 68 students provided satisfactory proofs, while in the control group it was only 2 of 68. In the control group, 22 students had not attempted any proof (in the experimental group 4). The other values are given in Table 4.3.

The feedback from the instructors was clearly in favor of the nonstandard approach. In the first part of the questionnaire, the instructors were asked to indicate their agreement or disagreement with certain statements (see Table 4.4).[6] In the second

---

[4] The comparability of the groups was documented by the *SAT mathematics ability scores* available for both groups (see [210], p. 372).

[5] The task is to correctly apply the definition of a limit at a point of discontinuity. According to Keisler's definition of a limit, it must be shown that $0 \neq \alpha \approx 0$ implies $f(2 + \alpha) \approx 4$ which results from the simple calculation $f(2 + \alpha) = (2 + \alpha)^2 = 4 + 4\alpha + \alpha^2 \approx 4$. According to the standard definition of a limit, we must instead do the following: For an arbitrary given null sequence (a_n), $a_n \neq 0$, and an arbitrary given $\epsilon > 0$, determine an n_0 such that $|f(2 + a_n) - 4| < \epsilon$ holds whenever $n \geq n_0$.

[6] In the original, further questions were asked, and the agreement or disagreement was differentiated according to two intensity levels. In this respect, the presentation in Table 4.4 is both an excerpt and a summary.

Table 4.3: Student responses to Question 3 ([210], p. 373).

	Control Group (68 students)	Experimental Group (68 students)
Did not attempt	22	4
Standard arguments		
satisfactory proof	2	
correct statements falling short of a proof	15	14
incorrect arguments	29	23
Nonstandard arguments		
satisfactory proof		25
incorrect arguments		2

Table 4.4: Instructor Questionnaire: Part One ([210], p. 374, excerpt and summary).

	agree	neutral	disagree
The students had a problem with accepting axioms for the hyperreal numbers.	4	1	7
Students seemed to find "infinitely small" a natural concept.	9	2	1
I think that a student who has had two semesters of nonstandard calculus will be at a disadvantage in a standard 3rd semester calculus course.	1	1	10
I am afraid that the introduction of infinitesimals left the students confused about the real numbers.	2	0	10

part, the instructors were asked which approach had advantages in relation to various aspects (see Table 4.5). According to the instructors' predominant assessment, the nonstandard approach has a cognitive advantage over the standard approach, especially for learning and understanding the basic concepts and due to the more intuitive proofs.

Sullivan summarizes that her investigation supports the thesis that Keisler's approach is indeed a viable alternative for teaching calculus. Fears that students learning calculus via this approach would be less able to master the basic skills are not justified.

Table 4.5: Instructor Questionnaire: Part Two ([210], p. 374). Possible responses: Standard (S), Nonstandard (NS), No Difference (ND).

Statement applies more to	S	NS	ND
Students learn the basic concepts of calculus more easily.		8	4
The students seem to be more "turned on."		5	7
The proofs are easier to explain and closer to intuition.	1	10	1
The students find it easier to ask their questions.		2	9
The students end up with a better understanding of the basic concepts of calculus.		5	7

The positive impressions were confirmed by Wattenberg on the basis of teaching experience at universities in Wisconsin and Massachusetts (see [225]).

More recently, Hernandez and Fernandez confirmed Sullivan's results with a study at the University of Puerto Rico in Rio Piedras and supplemented them with results on integral calculus [91], as did Ely at the University of Idaho [72]. Another relatively recent survey that also points in this direction was conducted at Bar-Ilan University (see Section 4.3.3). However, distorting influences in the studies cited, such as a different level of commitment in the teaching of standard and nonstandard analysis, cannot be completely ruled out. Likewise, it cannot be ruled out that the identified advantage of nonstandard analysis could be relativized by other measures in the teaching of calculus. On the other hand, I am not aware of any empirical studies that would explicitly refute the hypothesis of a cognitive advantage of the nonstandard approach or, conversely, prove a cognitive disadvantage.

There are voices that reject Keisler's axiomatic introduction of hyperreal numbers for fundamental reasons (see, for example, the review [36] by Errett Bishop). However, it should be noted that Bishop's judgment is based on a very specific, namely intuitionistic, foundational position (see also Section 6.1.2).

Cautiously formulated, it can be stated that the cognitive advantage hypothesis in favor of the nonstandard approach is not clearly proven, but is supported rather than refuted by the existing comparative studies.

4.3.2 The cognitive existence of infinitesimals

Most mathematicians today are willing to accept the existence of a mathematical object if its existence can be proven in ZFC. Some feel better about objects that can be constructed in ZF alone. Constructivists place even stricter requirements on the construction (see Section 5.1.5). For students (and even more so for pupils), however, a more intuitive understanding of existence can be assumed, which is neither linked to a proof of existence in ZF or ZFC nor to a constructivist proof of existence. Tall has coined the term *cognitive existence*, which is intended to express that concepts become part of an acceptable coherent structure in the student's mind ([216], p. 4). He then also speaks of a *belief in cognitive existence* or *cognitive belief* in existence.

In an introductory course on nonstandard analysis for third-year students, he investigated how the cognitive belief in the existence of number systems with infinitesimals (especially superreal and hyperreal numbers) developed during the course. In contrast to Sullivan, the students already had two years of experience in standard analysis.

In the course, the superreal numbers were not explicitly constructed, but introduced as power series that can be manipulated algebraically and visualized geometrically. The hyperreal numbers were first introduced axiomatically and then constructed using Zorn's lemma.[7] Immediately after this construction, the first survey was conducted (see Table 4.6), followed five weeks later, at the end of the course, by a second survey (see Table 4.7). The question in each case was: Do you consider the following as coherent mathematical ideas? The answer options were: definitely yes (1), fairly sure (2), neutral / no opinion (3), confused (4), fairly sure not (5), definitely not (6).

Table 4.6: Result of the survey at the beginning of the course ([216], p. 5).

N = 42	1	2	3	4	5	6
natural numbers	40	2	0	0	0	0
real numbers	39	3	0	0	0	0
complex numbers	32	8	1	0	1	0
infinitesimals	23	9	8	0	2	0
superreal numbers	18	12	8	2	2	0
hyperreal numbers	15	7	11	6	3	0

Table 4.7: Result of the survey at the end of the course ([216], p. 5).

N = 46	1	2	3	4	5	6
natural numbers	43	3	0	0	0	0
real numbers	39	5	1	0	0	0
superreal numbers	27	13	4	2	0	0
hyperreal numbers	15	20	5	4	2	0

Tall notes that the affirmative answers (categories 1 and 2) for infinitesimals in the first survey are even slightly higher than for superreal and hyperreal numbers and that the cognitive belief in the existence of superreal or hyperreal numbers has increased through the use of these number systems during the course. A further survey revealed that only 7 of the 46 course participants considered it a flaw that the superreal numbers had not been formally constructed. Ten were positively unhappy with the use of Zorn's lemma in the construction of the hyperreal numbers, although 22 considered the construction to be essential. Tall interprets this to mean that the construction of numbers is more likely to be considered necessary when there is less cognitive belief in their existence (cf. [216], p. 6).

Robert Ely confirmed in his dissertation [71] that many students have robust conceptions of a real number line that include infinitesimal and infinitely large quantities and distances. Furthermore, 31 % of the calculus students he surveyed "claimed consistently over multiple questionnaire items that there exist infinitely small numbers and/or

---

[7] According to Tall, the students had little or no experience with Zorn's lemma up to this point.

distances" ([71], cited in [70], p. 139). In a detailed case study with a student (Sarah), he found numerous similarities between Sarah's and Leibniz's ideas, which were later formalized in Robinson's nonstandard analysis (see, for example, the tabular overview in [70], pp. 140–141). Ely concludes:

> These similarities suggest that these student conceptions are not mere misconceptions, but are nonstandard conceptions, pieces of knowledge that could be built into a system of real numbers proven to be as mathematically consistent and powerful as the standard system. This provides a new perspective on students' "struggles" with the real numbers, and adds to the discussion about the relationship between student conceptions and historical conceptions by focusing on mechanisms for maintaining cognitive and mathematical consistency ([70], p. 117).

The evaluation in [22] on the question "$0,\overline{9} < 1$ or $0,\overline{9} = 1$?" spoke also very clearly in favor of the idea of the infinitely small. In total, 256 secondary school students in grades 7 to 12, 50 mathematics students (after the third semester), and 51 student teachers of various subjects were surveyed and 50 % of the mathematics students were in favor of $0,\overline{9} < 1$. Among student teachers of other subjects, it was even over 90 %, and among secondary school students over 72 %. Even three semesters of ε–δ analysis apparently could not eradicate the idea of infinitely small differences between numbers in half of the mathematics students.

4.3.3 A-track vs. B-track

Mikhail G. Katz and Luie Polev from Bar-Ilan University in Ramat Gan, Israel, have taught calculus to approximately 120 freshmen of the 2014/15 class using Keisler's approach. First, the basic concepts were defined with infinitesimals, and only then were the ε–δ definitions given. This approach was maintained in the following two classes until the publication in [114]. Katz and Polev call the first path *B-track* and the second path *A-track*.[8]

At the end of the course, the students were asked which definitions they found more helpful in understanding the basic concepts of calculus, specifically *continuity*, *uniform continuity*, and *convergence*. To this end, the students were asked to evaluate the statement "the definition helped me understand the concept" according to the following scheme: (1) agree strongly; (2) agree; (3) undecided; (4) disagree; and (5) disagree strongly.

In total, 84 students took part in the poll. The result is summarized in Table 4.8 (responses (1) and (2) counted as "found helpful"). Katz and Polev comment on the result as follows:

---

[8] The letters A and B are meant to recall Archimedes and Bernoulli, respectively, since the A-track uses an Archimedean continuum, while the B-track uses infinitesimals, like Bernoulli.

Table 4.8: Result of the poll by Katz and Polev ([114], pp. 94–95). The numbers in parentheses refer to the students who were able to define the concepts correctly.

For understanding the concept of	continuity	uniform continuity	convergence
the A-track definition found helpful:	10 % (9 %)	21 % (24 %)	10 % (13 %)
the B-track definition found helpful:	69 % (75 %)	74 % (80 %)	62 % (70 %)

To summarize, what we tried to do in the course is to impart to the students the fundamental concepts of the calculus in a way that is the least painful to the students, while making sure that they have the necessary background in the ε–δ techniques to continue in the second semester course taught via EDC [Epsilon–Delta Calculus]. The results of the poll suggest that starting with the intuitive B-track definitions succeeds in this sense. Once the students understand the basic concepts via their intuitive B-track formulations, they are able to relate more easily to the A-track paraphrases of the definitions ([114], p. 94).

The questionnaire also contained a control question asking the students to prove $\lim_{x \to 2}(x + 5) = 7$ in two different ways, once via A-track and once via B-track.[9] All in all, 98 % attempted a proof via B-track (of which 85 % were successful), 71 % attempted a proof via A-track (of which 20 % were successful).

4.3.4 Is there resistance?

In view of the reported positive experiences with experimental nonstandard courses in calculus, the question arises as to why this approach is so little used in teaching. Tall sees a connection between the effort invested in ε–δ analysis, on the one hand, and the appreciation of this approach or the resistance to alternative approaches, on the other. During his study on the cognitive existence of infinitesimal numbers (see Section 4.3.2), he observed this correlation in interviews with students and also suspects it among teachers.

> When several students, representing a cross-section of all abilities, were interviewed in depth after the course, it became clear that their heavy investment in ε–δ analysis made them have a high regard for it, even though it still presented them with technical difficulties.
> The vast majority of university teachers have a similar investment, so a cultural resistance to nonstandard analysis is only natural ([216], p. 6).

Accordingly, standard analysis in teaching would be a self-perpetuating system, and any alternative approach would have virtually no chance of success. In fact, neither Keisler at Wisconsin nor Katz at Bar-Ilan University have managed to establish their experimental analysis courses permanently in the mathematics department.

---

9 The different solution paths are analogous to those in footnote 5.

Keisler began teaching introductory calculus courses using hyperreal numbers in 1969. His resulting 1976 textbook *Elementary Calculus* for a three-semester calculus course was used for about 20 years at the University of Wisconsin Madison. During this time Keisler was able to recruit about nine other members of the math department to teach the course. However, since the number of volunteers eventually ran out, the course was not sustainable, as Rebecca Vinsonhaler reports from her personal communication with Keisler in 2014 [223]. According to Vinsonhaler, Keisler gave the following reasons for the difficulties in finding volunteer teachers for the course (cf. [223], pp. 271–272):

- It is much easier for a teacher to use the approach they are familiar with than to do extra work to learn a new approach (in particular since calculus is considered a "service course").
- Many professors were uncomfortable with the material and therefore not confident in their ability to teach it. The lack of familiarity with mathematical logic also plays a role.
- Math departments think it is dangerous to tamper with their calculus course. If the other departments did not approve of the experiments and decided to offer their own calculus courses, this would lead to a loss of control and possibly even to a large decrease in the number of positions within mathematics departments.
- The volunteer teachers who taught Keisler's course sometimes faced hostility from their colleagues.

Mikhail Katz also reports a certain hostility within the mathematics department towards the nonstandard approach, despite positive feedback from students:

> We taught using the infinitesimal method for 5 years in the computer science department, and trained close to 1000 students. We also taught in the mathematics department for 2 years where the classes were considerably smaller: on the order of 40 students in each class. The students were very satisfied but the approach generated a considerable amount of hostility among the faculty and was abandoned last year. This year a new chairman came in the computer science department who seems more favorably inclined. We did opinion surveys among computer science students who took our courses and they are overwhelmingly in favor of the infinitesimal approach (they are familiar with both approaches, both because we taught epsilon–delta in our course, and also because the follow-up second semester course was pure epsilon–delta). This seems to have made an impression on the current chair. At any rate it remains to be seen if this is ever reinstated. In the math department there are also a couple of people who are favorably inclined but the higher-ups ... oppose it (Katz, personal communication, September 30, 2020).

According to Katz, one argument put forward by the other side was that "we have got to be able to give the students a complete and satisfying answer to the question: 'what is a number?'"

Two things are interesting here. Firstly, the mathematics department's argument against the nonstandard approach: Apparently, it is assumed that an axiomatic introduction of real numbers will fully and satisfactorily answer the students' question of what numbers are, whereas an axiomatic introduction of hyperreal numbers will not. Secondly, computer scientists (at least at Bar-Ilan University) seem to have fewer reservations about nonstandard analysis than mathematicians (contrary to the concern that Keisler suspected in the mathematics department regarding the other departments). One explanation could be that for computer scientists the real numbers (as actual infinities) seem just as far removed from reality as the hyperreal numbers, while for many mathematicians the real numbers *are* virtually reality (see also the discussion in Chapter 5).

From an educational point of view, it is important to note that in addition to the infinitesimal method, the epsilon–delta technique was also taught in the courses offered. The use of infinitesimals in the calculus courses was therefore not about replacing the limit concept, but about making it easier to get started. Keisler's *Elementary Calculus* also deals with the epsilon–delta definition of limit (Section 5.8 in [120]).

4.4 A survey among analysis teachers

In April 2018, a survey on the use of nonstandard analysis in teaching was conducted at 66 German universities with a mathematics department as part of the author's dissertation project. The aim of the survey was to find out

1. whether elements or methods of nonstandard analysis are taken into account in Analysis I/II lectures,
2. whether, in the opinion of the lecturers, nonstandard analysis could be used sensibly in teaching (possibly in addition to the standard lectures, for example, in an undergraduate seminar),
3. what the main reasons for the respective assessment are.

The addressees of the survey were the lecturers of the introductory courses on analysis (usually Analysis I) in the previous winter semester. The questions were asked by email without predefined answer categories and the responses were analyzed using Mayring's *Qualitative Content Analysis* [153, 151, 152]. We were able to analyze 50 responses to the first question and 29 responses to the second and third questions. We summarize the results here. Further details on the survey and its analysis can be found in [124].

4.4.1 Responses regarding current practice

The results of the survey confirmed the picture that emerged from the textbook analysis (see Section 4.1): Nonstandard analysis plays virtually no role in university teaching (in Germany). None of the 50 lecturers who took part in the survey used elements or methods of nonstandard analysis in their Analysis I or II lectures. Two people stated that they had already held an undergraduate seminar (in Germany called *proseminar*) on nonstandard analysis. One person was inspired by the survey to possibly offer such an undergraduate seminar in the future.

4.4.2 The assessment of nonstandard analysis (from the teachers' perspective)

The responses to the assessment of the use of nonstandard analysis in teaching can be summarized as follows:
- 10 people (34 %) rated the use in teaching exclusively negatively (generally *not suitable* or *less suitable* for all types of courses).
- 4 people (14 %) considered the use to be *possible* or *suitable* at most in special courses in advanced studies.
- 6 people (21 %) were at least neutral about complementary use (*possible* for complementary courses).
- 9 people (31 %) were positive about the complementary use (*well suited* for complementary courses); 2 of these people (7 % of all) were also positive about the use in the lectures (*well suited* for the lectures).

The percentages refer to the 29 responses that could be evaluated for the assessment question.

4.4.3 The teachers' arguments

The arguments for and against the use of nonstandard analysis in teaching mentioned by the teachers are summarized in Table 4.9. The number of interviews in which the respective argument was mentioned (out of a total of 29 interviews) is given in brackets. The categories were formed inductively on the basis of the free text responses.

4.4.4 Reasons for a negative attitude towards nonstandard analysis in teaching

Conflicts with habits of thought or values are possible causes for a negative attitude towards the object that triggers these conflicts. When interpreting the results in Table 4.9 with regard to possible reasons for a negative attitude towards nonstandard analysis, it therefore makes sense to focus on the categories that suggest such conflicts and less on the categories that point to unfavorable conditions. I see the following starting points here:

Table 4.9: Reasons given by teachers for (+) or against (−) the use of nonstandard analysis in teaching.

	Main category	Category (number of mentions)
+	Cognitive advantage	Promoting understanding (2) More elegant and intuitive proofs (1) Building intuition (1)
+	Repetition of material	Repetition of material in complementary courses (1)
+	Awareness of the foundations of mathematics	Promoting awareness of the foundations of mathematics (1)
−	Unfavorable conditions	Deviation from content specifications (3) Few suitable textbooks (1) Lack of human resources (1) Lack of competence among teachers (1) Lack of time in the basic lectures (6)
−	Excessive demands	Lack of prior knowledge among students (3) Excessive demands on students (2) Confusion among students (4) High degree of abstraction (2)
−	Little relevance	Little relevance for mathematics (14)
−	Personal reasons	Little relevance to my own field of research (1)
−	Lack of benefit	Little added value compared to standard analysis (5) Of no use in the further course of studies (4)

Excessive demands

Assumption:	Nonstandard analysis cannot be taught with the necessary rigor in beginner courses (due to a lack of prior knowledge among students).
Consequence:	Rejection due to the conflict with the value of "mathematical rigor" in university teaching.
Assumption:	Nonstandard analysis is too difficult for beginners.
Consequence:	Rejection due to conflict with recognized educational principles (do not confuse or overwhelm, no inappropriately high abstraction right at the beginning).

Little relevance

Assumption:	Nonstandard analysis is of little relevance to mathematics.
Consequence:	Rejection due to the conflict with the desire to teach something relevant.

Lack of benefit

Assumption:	Nonstandard analysis is of no use in the further course of studies (as it is no longer needed) and has no added value compared to standard analysis.
Consequence:	Rejection due to the conflict with the desire to teach something useful.

In order to examine whether the above assumptions are justified judgments or rather prejudices, we will examine the following questions:
- How difficult is nonstandard analysis?
- How relevant is nonstandard analysis?
- What are the benefits of nonstandard analysis?

4.4.5 How difficult is nonstandard analysis?

The overview in Chapter 3 has shown that nonstandard analysis is a challenging topic. It should be widely agreed that neither Robinson's model-theoretical argumentation, nor Luxemburg's ultrafilter construction, nor Nelson's axioms of internal set theory or related axiom systems can be taught in first-year analysis lectures. The concepts required for this are indeed too difficult or not available for beginners. However, they are also not needed if one only wants to counteract potential difficulties in understanding the basic concepts of calculus. In Section 4.2, various reduced programs were presented with which nonstandard analysis can also be made accessible to beginners. An elementary axiomatic approach using an extension of the language was discussed in detail in Chapter 2.

The approach according to Henle [89] requires the least changes compared to the standard lectures, as it still operates directly with the real-valued sequences, but at the same time uses concepts and methods of nonstandard analysis. Additional axioms or a construction of new numbers by means of equivalence classes are not necessary here. Laugwitz had already proceeded in this way in his beginners' lectures (cf. [130], p. 242).[10]

An approach based on the *elementary extension principle* (hereinafter also referred to as *extension principle* for short), which is based on Keisler and Henle/Kleinberg, is also very efficient. In its short form, the principle reads:
- There is an elementary extension of the structure of real numbers.

In more detail it reads:

Elementary extension principle. *There is a proper extension* $^*\mathbb{R}$ *of* $\mathbb{R}$ *and a mapping* $*$ *that*
1. *assigns to each n-ary relation R on* $\mathbb{R}$ *an n-ary relation* $^*R$ *on* $^*\mathbb{R}$,
2. *assigns to each n-ary function* $f: D \to \mathbb{R}$ *(with* $D \subseteq \mathbb{R}^n$*) an n-ary function* $^*f: {}^*D \to {}^*\mathbb{R}$.
3. *satisfies the following transfer principle: If* φ *is a first-order statement that holds in* $\mathbb{R}$*, then the statement* $^*\varphi$ *holds in* $^*\mathbb{R}$*, where* $^*\varphi$ *arises from* φ *by replacing all functions and relations by their images under the mapping* $*$.

---

[10] Further lectures in the style of his book "Zahlen und Kontinuum", which deal with the topic of infinitesimal mathematics in more detail, were held at intervals of about four years for intermediate students (cf. [130], p. 242).

First-order statements do not have to be formally defined. It is sufficient to characterize them informally by the restriction that quantifications are only permitted over number variables.

If the extension principle is motivated by Leibniz's guiding ideas on infinitesimal calculus (e.g., analogous to the historical starting points in Section 2.1) and explained using a few examples, it does not seem too difficult for calculus courses, as the experiences described in Section 4.3 confirm. Basically, it is even quite easy to understand if one takes the real numbers as given.[11] The decisive advantage of this approach is that with a single additional principle, the extension principle, nonstandard proofs can be carried out with the same rigor as standard proofs.

It is a different question whether the extension principle is acceptable as an additional axiom in an introductory course without having concrete examples of nonstandard numbers. In this respect, the extension $^*\mathbb{R} \supset \mathbb{R}$ differs from the extension $\mathbb{C} \supset \mathbb{R}$, where one *constructs* $\mathbb{C}$ as $\mathbb{R}^2$ (and then embeds $\mathbb{R}$).

It is true that in a beginners' lecture the axioms of the real numbers must also be accepted without seeing a construction of the real numbers. However, there you have the advantage of finding the real numbers as the "familiar infinite decimal fractions".[12] So we ask once again: Is the extension principle acceptable as an axiom in an introductory course?

The following points can be made here. First, in Sullivan's study (see Section 4.3.1), the students (according to the teachers' impression) had no problems accepting the axioms of hyperreal numbers (essentially the extension principle).[13] Second, the extension principle is a postulate that – like the axioms of the real numbers – is mathematically justified (since there are corresponding models) and that can serve students to open up the historically guided and intuitive conceptual world of calculus on a precise basis. In the experimental courses with Keisler's approach described above, students are then also given Weierstrass's definitions (which do without the extension principle), but still have the cognitive bridge to the more intuitive nonstandard concepts. Third, finally, the extension principle is in a sense a consistent generalization (and thus reinforcement) of the requirement that $^*\mathbb{R}$ should be an ordered extension field of $\mathbb{R}$. The generalization in the extension principle consists in the requirement that the mapping $*$ assigns an extension not only to the relation $<$ and the functions $+$ and $\cdot$, but to every finitary relation

---

[11] Formally, the extension principle is an axiom schema in a language with an uncountable number of symbols (see Section 3.3.3), but axioms are not dealt with at this formal level in the beginners' lectures.

[12] The term is in quotation marks because infinite decimal fractions turn out to be less familiar on closer analysis than one might think, which is due to the fundamental problem of the actual infinite (see the discussion in Chapter 5 (especially in Sections 5.2, 5.4, and 5.5)). For a criticism from a pedagogical perspective of the introduction of real numbers as infinite decimal fractions, see also [23] and [24].

[13] In Tall's experiment (see Section 4.3.2), about half of the students stated that they considered the construction of hyperreal numbers to be essential, but the situation is not comparable because the students there were third-year students.

and every finitary function on ℝ, and that the transfer principle applies not only to the field and order axioms, but to all first-order statements.

Overall, it should be noted that nonstandard methods are not necessarily too difficult to be used in beginners' lectures. According to previous teaching experience, the programs available for this do not lead to confusion among students. The feedback from teachers and students described in the Sections 4.3.1 and 4.3.3 tends to point in the opposite direction: the inclusion of nonstandard methods was perceived by students as promoting motivation and understanding. According to the experiences reported in Section 4.3, the ability to connect to standard courses and to apply calculus correctly and solve standard examination tasks was not impaired by the nonstandard introduction.

The contention that students are overburdened by excessive abstraction also does not apply to the reduced programs discussed here. On the contrary, the availability of infinitely small and infinitely large quantities opens up completely new possibilities for visualization. The instrument of infinite magnification or reduction through corresponding "microscopes" or "telescopes" is used intensively by Keisler, Tall, Deledicq, and Diener, as well as other textbook authors, and is seen as a significant advantage of a nonstandard introduction to calculus. The extent to which this instrument is effective was examined in [125].

4.4.6 How relevant is nonstandard analysis?

In the preface to their 1994 textbook, Landers and Rogge write about the importance of nonstandard mathematics: "Nonstandard mathematics has experienced a great upturn in recent decades. It has influenced and stimulated developments in the most diverse fields" ([127], p. V).[14] The authors go on to explain:

> It has been shown that nonstandard methods are a powerful tool for the treatment of mathematical problems. Since Robinson, nonstandard methods have been used to prove known results in a more transparent and natural way, on the one hand, and to gain new mathematical insights and solve open problems in classical mathematics, on the other hand. Nonstandard methods have been used very successfully in topology, functional analysis, stochastics, mathematical physics, and mathematical economics. In the applied sciences in particular, it has been shown that the nonstandard domain *ℝ is often better suited for modeling than the classical domain ℝ of real numbers ([127], p. 2).[15]

---

[14] In the original German: "Die Nichtstandard-Mathematik hat in den letzten Jahrzehnten einen großen Aufschwung erfahren. Sie hat die Entwicklungen in den verschiedenartigsten Gebieten beeinflusst und befruchtet."

[15] In the original German: "Es hat sich gezeigt, daß Nichtstandard-Methoden ein mächtiges Instrument zur Behandlung von mathematischen Fragestellungen sind. Nichtstandard-Methoden wurden seit Robinson dazu eingesetzt, um einerseits bekannte Ergebnisse durchsichtiger und natürlicher zu beweisen und andererseits neue mathematische Einsichten zu gewinnen sowie offene Probleme der klassischen Mathematik zu lösen. Sehr erfolgreich eingesetzt wurden Nichtstandard-Methoden bisher in der Topolo-

Landers and Rogge themselves focus on topology and stochastics. Numerous examples of applications in various fields can be found, for example, in
- *Nonstandard methods in stochastic analysis and mathematical physics* by Albeverio, Høegh-Krohn, Fenstad, and Lindstrøm [2],
- *Nonstandard Analysis: Theory and Applications* by Arkery, Cutland, and Henson [7],
- *Applied nonstandard analysis* by Davis [57],
- *Nonstandard Analysis in Practice* by Diener and Diener [61],
- *Nonstandard analysis for the working mathematician* by Loeb and Wolff [141],
- *Nonstandard Analysis: A Practical Guide with Applications* by Lutz and Goze [144],
- *Nonstandard Analysis* by Väth [224].

For applications in economics in particular, see also Anderson's *Infinitesimal methods in mathematical economics* [3]. The book by Loeb and Wolff also contains a section on economics. Väth, who is particularly interested in topology and functional analysis, sees the strength of nonstandard analysis in the fact that "it provides a machinery which enables one to describe 'explicitly' mathematical concepts, which by standard methods can only be described 'implicitly' and in a cumbersome way," for example, Hahn–Banach limits (see [224], p. vii, quotation marks in the original).

Prominent successes of nonstandard analysis include the solution of an *invariant subspace problem* by the theorem of Bernstein and Robinson (see [33]) and the significant simplification of Gleason's solution of Hilbert's fifth problem (proof that every locally Euclidean group is a Lie group, see [97] and [220]).

If we look at the number of publications classified as "26E35 Nonstandard analysis" (according to the MSC2020 Mathematics Subject Classification System) in zbMATH Open, we can see that nonstandard analysis has been an active field of research since the 1960s and still is (with particular peaks in the 1980s and 1990s). Figures 4.1 to 4.6 show the distribution of publications from 1963 to 2022 with the following classification codes:[16]
- 26E35 Nonstandard analysis
- 03H05 Nonstandard models in mathematics
- 03H10 Other applications of nonstandard models (economics, physics, etc.)
- 03H15 Nonstandard models of arithmetic
- 28E05 Nonstandard measure theory
- 54J05 Nonstandard topology

It should be noted that the documents are often assigned to several classification codes (for example, 26E35 and 54J05).

---

gie, Funktionalanalysis, Stochastik sowie in der Mathematischen Physik und der Mathematischen Ökonomie. Gerade in den angewandten Wissenschaften hat sich gezeigt, daß der Nichtstandard-Bereich *ℝ zur Modellbildung häufig besser geeignet ist als der klassische Bereich ℝ der reellen Zahlen."

16 The numbers are taken from the website https://www.zbmath.org/ (visited on April 2, 2023). Figures: Own graphical representation.

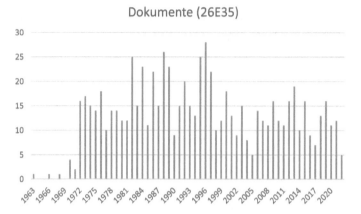

Figure 4.1: Number of published documents with MSC 26E35 (Nonstandard analysis).

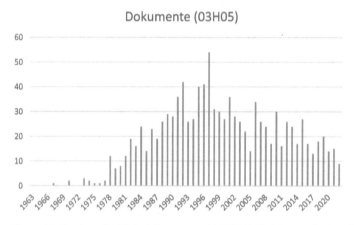

Figure 4.2: Number of published documents with MSC 03H05 (Nonstandard models in mathematics).

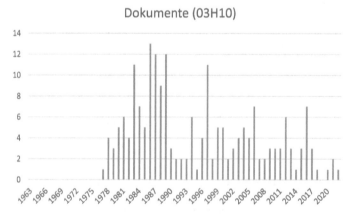

Figure 4.3: Number of published documents with MSC 03H10 (Other applications of nonstandard models (economics, physics, etc.)).

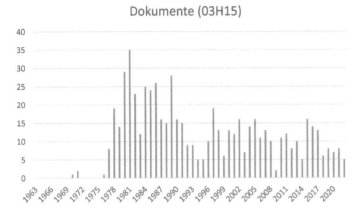

Figure 4.4: Number of published documents with MSC 03H15 (Nonstandard models of arithmetic).

Figure 4.5: Number of published documents with MSC 28E05 (Nonstandard measure theory).

Figure 4.6: Number of published documents with MSC 54J05 (Nonstandard topology).

4.4.7 What are the benefits of nonstandard analysis?

Limits are one of the basic tools of mathematics. The literature on calculus is written in the language of limits (according to Weierstrass' definition). Anyone wishing to access this literature must therefore be familiar with the language of limits. This inevitably leads to the following consequence:
- If you are aiming for a degree in mathematics, you must be able to understand and apply Weierstrass' limit concept. Definitions with three alternating quantifiers should not be an obstacle.

And it is certainly true that
- Weierstrass' limit concept is not difficult to understand compared to almost everything else that comes along in mathematics studies.

So is there any reason at all to think about simplifying the introduction to calculus? Or is the limit concept simply an early litmus test for your suitability to study mathematics?

As undeniable as the fact that the limit concept is central to calculus is the fact that many first-year students struggle with it.[17] But not everyone who has difficulties at the beginning is unsuitable for studying mathematics. The difficulties often lie generally in the transition from elementary, rather informal school mathematics to advanced, axiomatic, and more formal university mathematics. Anything that makes this transition easier for students is therefore welcome. Tall describes the transition from elementary to advanced mathematics as follows:

> The move from elementary to advanced mathematical thinking involves a significant transition: from *describing* to *defining*, from *convincing* to *proving* in a logical manner based on those definitions. This transition requires a cognitive reconstruction which is seen during the university students' initial struggle with formal abstractions as they tackle the first year of university. It is the transition from the *coherence* of elementary mathematics to the *consequence* of advanced mathematics, based on abstract entities which the individual must construct through deductions from formal definitions ([214], p. 20, emphasis in the original).

The reasons for the difficulties of this transition are manifold (see, for example, the sources in footnote 17). With regard to possible advantages of nonstandard analysis, the following aspects appear worth emphasizing.
- The challenge of understanding more complex logical formulas, especially those with nested quantifiers (see [194]).
- A discrepancy between ideas that students associate with a concept (*concept image*) and the *concept definition* [218].

---

17 See, for example, [35, 163, 164, 211, 212, 218, 215].

Formulas with up to three nested quantifiers play a role right at the beginning (when defining limits), especially in calculus. Their correct interpretation is crucial for understanding and applying the concepts defined in this way. Subtleties such as the order of the quantifiers and the associated dependence of the variables on each other are relevant (for example, when differentiating between continuity and uniform continuity). This is where the nonstandard definitions offer advantages due to their reduced complexity.

One difficulty in understanding the limit concept lies in the "unencapsulated definition" [54]. The dialogic in the definition $\forall \varepsilon > 0 \ \exists N \in \mathbb{N} \ldots$ ("You give me any ε, and I will find a suitable N") suggests the idea of a process in which a suitable N is to be specified for a sequence of ever smaller ε. For students, therefore, "limit is often seen in terms of the potential infinity of the on-going process rather than the fixed limit that can be calculated to any desired accuracy" ([43], p. 4).

Oehrtman has cataloged the "metaphors" that students employ as they attempt to explain the meaning of statements involving limits [163]. These include the metaphors *collapse* (the process collapses into the limit at some point), *proximity* (if x is close to y, then $f(x)$ is close to $f(y)$), *infinity as number* (infinity, which is used in equations or inequalities or as an argument in functions), *physical limitation* (assumption of a smallest positive number) and *approximation* (cf. [43], p. 7).

The assumption of a smallest positive number (*physical limitation*) is compatible neither with standard nor with nonstandard analysis. The metaphor *approximation* can be interpreted both in standard analysis (as an arbitrary approximation) and in nonstandard analysis (as an infinite approximation). The other metaphors have a more direct and precise interpretation in nonstandard analysis: *infinity as number* as unlimited nonstandard numbers, *proximity* as infinitesimal neighborhood, and *collapse* as the transition to the standard part. This fits Ely's finding that students have robust conceptions of a number line including infinitesimal and infinite distances (see Section 4.3.2) and the results of Katz and Polev, according to which the nonstandard introduction was found to be helpful for understanding the basic concepts and also had a positive effect on the understanding of the standard definitions (see Section 4.3.3).

Another aspect to consider is that by no means all students who take calculus courses are aiming for a degree in mathematics. In natural science or engineering courses, it is common to argue informally with infinitesimals rather than with strict "epsilonics". Students of these subjects are likely to find a nonstandard introduction to calculus convenient.[18]

In Section 4.3.1, the hypothesis of the cognitive advantage of infinitesimal mathematics was discussed, for which there is some evidence in the literature and (as far as

---

[18] A historical note: As early as the nineteenth century, there was resistance to an excessive formalization of calculus on the basis of Weierstrass's limit concept, as it was considered unsuitable for application-oriented engineering courses (see [175]).

I know) no explicit refutation. Nonstandard analysis makes an *offer* for teaching that should at least be considered. If one follows the cognitive advantage hypothesis, this is a didactic benefit and added value of nonstandard analysis. The point is not to abolish or replace the limit concept, but about making it easier to get started with calculus. Nonstandard analysis supports and complements standard methods. This is true not only from a didactic point of view, but also mathematically (see Section 4.4.6).

A further benefit of nonstandard methods in teaching therefore lies in broadening horizons with regard to methods in mathematics. If students learn early on that there is nothing reprehensible about calculating with infinitely small and infinitely large numbers, this may awaken their interest in studying nonstandard methods in more detail later on and including them in their range of methods when specializing in advanced studies, for example, in functional analysis, topology, or stochastics. In this respect, it cannot be claimed that nonstandard analysis has no benefit for later studies.

4.4.8 Are there other reasons for a negative attitude?

The reasons discussed so far for a negative attitude towards nonstandard analysis in teaching were essentially due to prejudices that do not stand up to closer scrutiny. However, a negative attitude can also be caused by habits of thought that are linked to certain consciously or unconsciously adopted philosophical positions. In Section 4.3.4 we had already identified the argument from the mathematics department of Bar-Ilan University against the experimental analysis courses ("We have got to be able to give the students a complete and satisfying answer to the question: 'what is a number?'") as an indication of reasons for rejection of this kind. The fact that "little relevance for mathematics" was mentioned most frequently in the survey as a reason against the use of nonstandard analysis in teaching could also be an indication of philosophically based rejection because relevance is always judged from a certain perspective on mathematics.

It therefore seems worthwhile to work out the role of nonstandard analysis for the philosophy of mathematics in more detail in order to be able to identify and evaluate habits of thought as possible causes for a rejection of nonstandard analysis in teaching. This is done in the following Chapter 5.

5 Philosophical questions and mathematical challenges

5.1 From the foundations of mathematics

5.1.1 A look at history

Is mathematics conceivable without philosophy? The early advanced civilizations in Egypt and Sumer had extensive mathematical knowledge without any evidence of philosophical engagement with mathematics having come down to us. Even today, it is possible to study mathematics or carry out mathematical research without worrying about philosophical questions. Philosophy comes into play when we stop, take a step back, and reflect on what we are doing when we do mathematics.

Philosophy and mathematics have been closely intertwined since Greek antiquity at the latest. Original philosophical questions about the nature of things and the possibilities of knowledge arose in a special way in relation to mathematics, whose objects seemed to be detached from immediate sensory experience, on the one hand, and realized in many ways in the world of experience, on the other.

For Plato, the realm of mathematics was part of an ideal reality that is recognizable to humans based on their intuition, a kind of memory of the soul. The philosophical position of *mathematical Platonism* or *metaphysical realism* derived from this remained predominant until modern times and is still attractive today. For over two thousand years, Euclidean geometry was regarded as a prime example of certain knowledge and its axiomatic method, the *more geometrico*, as exemplary for a rigorous acquisition of knowledge that is beyond any doubt.

The development of algebra led to a multiple expansion of the concept of a number and thus to the question of the ontological status of the new numbers. To what extent could negative, irrational, imaginary numbers be regarded as *existent* or as numbers at all? The infinitesimal quantities that analysis came up with seemed to be even more problematic. Could infinitely small changes in a quantity, infinitely distant points or infinitieth terms of a sequence be used in proofs? Did such things exist? In view of the Aristotelian prohibition of actual infinities, this seemed more than questionable and in any case far removed from the rigor of a *more geometrico*.

According to Bedürftig and Murawski, we can speak of a philosophy of mathematics as a philosophical subfield from the nineteenth century onwards with the attempts to justify the analysis ([26], p. 397). The undeniable success of the infinitesimal calculus founded by Newton and Leibniz, on the one hand, and the perceived insufficient justification of its methods, on the other, made a philosophical examination of this new mathematics indispensable. The work of Cantor, Dedekind, and Weierstrass succeeded in reconstructing calculus without infinitesimal quantities. However, the price for this was to admit actual infinity in the form of infinite sets, which led to new criticism and new uncertainty. With the set-theoretical definition of the real numbers and the arithmetization of the continuum, intuition was replaced as the valid basis of mathematics by an abstract set concept. Already at the beginning of the nineteenth century, trust in geometric intuition as a reliable source of knowledge had been shaken by the discovery of non-Euclidean geometries. Axioms forfeited their claim to be evident, unquestionable truths and could later, in formalism, be regarded as mere agreements.

At the beginning of the twentieth century, mathematicians took their foundations into their own hands, so to speak. Mathematical logic and axiomatic set theory evolved; in the foundational crisis triggered by antinomies in set theory, the classical foundational positions of logicism, formalism, and intuitionism emerged. Mathematical logic, with its subdisciplines of proof and model theory, provided insights that in turn had an impact on philosophy. Gödel's incompleteness theorems are considered a milestone and a certain caesura, as they show fundamental limits of the axiomatic method in formal systems. The failure of Hilbert's program in its original form (as an attempt to finitistically justify infinitistic mathematics and especially set theory) as well as the inevitable incompleteness of formal theories (which are consistent and recursively axiomatizable and include arithmetic) led to a certain disillusionment. The undecidability of relatively simple but intuitively relevant statements, such as the continuum hypothesis, within the framework of usual set theory has led set theorists to search to this day for suitable extensions of the axiom system that bring about a decision in an intuitively plausible way.[1]

Everyday mathematical life goes on, undisturbed by such foundational problems. With mathematical logic and axiomatic set theory, the mathematical community has created a foundation that is considered suitable and sustainable by the vast majority. It forms (mostly unspoken) the accepted framework for mathematical work.

However, such a broad consensus on the mathematical foundations should not obscure the fact that the fundamental philosophical questions about the nature of mathematical objects and our ability to know something about them remain unanswered.

5.1.2 Is there right and wrong mathematics?

As we know mathematics from school or university, it often appears to us as absolute and objective, as a collection of definitions, theorems, and proofs, and of procedures for solving certain problems, as an essentially cumulative discipline that has accumulated knowledge over thousands of years, which is still being expanded and developed, but which remains stable in its stock. Existing concepts and theories can be generalized or

---

[1] The research program "V = ultimate L" by W. H. Woodin, for example, would (if successful) positively decide the continuum hypothesis (see [229]).

new connections can be discovered, completely new subdisciplines with new concepts can emerge, but what has once been proven to be true remains true for all time and was already true before the discovery of the proof. Mathematical truths are eternal and above all objective, i. e., free from opinions and personal preferences. Experts should always be able to agree on what is a valid definition or what is a valid proof.

Anyone who has studied the history or philosophy of mathematics knows that this ideal is not true. Mathematics is not (exclusively) cumulative, but evolutionary. And it is not (completely) objective, but embedded in a philosophical canon of values that (consciously or unconsciously) sets the framework within which mathematics takes place. Bedürftig and Murawski write:

> Finally, notice that the philosophy of mathematics that attempts to consider the phenomenon of mathematics as a science and to determine and discuss its foundations, describes, on the one hand, mathematics as it is done and developed and, on the other hand, should fix, reconsider, and evaluate methodological norms. Hence philosophy of mathematics has both descriptive and normative character ([26], p. 399).

Philosophy of mathematics therefore not only describes what mathematics *is* and how it *is* practiced, but also what it *should* be and how it *should* be practiced. Put simply, it distinguishes between good and bad mathematics or between right and wrong mathematics (in a normative sense).

Naturally, such an assessment is neither uniform nor final, but subject to historical change and at all times a struggle between different positions. In this respect, it is clear that fundamental philosophical convictions can also be the source of resistance to certain parts of mathematics.

The following questions may suffice as examples: Is constructive mathematics better than inconstructive, concrete better than abstract, applicable better than purely theoretical, finitistic better than infinitistic? Is the axiom of choice acceptable? Do implicit definitions define anything? Are predicative definitions better than impredicative? Are direct proofs better than indirect?

5.1.3 Fundamental questions of the philosophy of mathematics

The fundamental questions of the philosophy of mathematics can be roughly divided into three areas.

Ontological questions: In what way do mathematical objects exist? Is mathematics discovered or invented?

Epistemological questions: In what sense can we know something in mathematics? What does truth mean in mathematics? How do we gain knowledge? How can we justify the methods used to gain knowledge? What are the limits of knowledge?

Questions on applicability: What is the relationship between mathematics and reality? Why are the results of mathematics applicable to the world?

Depending on the philosophical position, these questions are answered very differently. The diversity of positions cannot and need not be presented here in their entirety (for an overview, see [26]). With realism (Section 5.1.4), constructivism (Section 5.1.5), and formalism (Section 5.1.6), we pick out three positions that sufficiently cover the spectrum for the following discussion.[2] The section on formalism is followed by sections on reverse mathematics (which is relevant in connection with Hilbert's program) and mathematical practice.

According to the fundamental questions of the philosophy of mathematics, the preference for standard analysis or the rejection of nonstandard analysis can be discussed under ontological, epistemological, and application-related aspects. With regard to the concept of number, one can ask: Are nonstandard numbers just as *real* as standard numbers? Can we know anything about nonstandard numbers just as reliably as about standard numbers? Can statements about nonstandard numbers have any *meaning* for reality? We will address these questions in the next chapter in the Sections 6.1 to 6.3.

Before that, we dedicate a separate section to infinity (Section 5.2), the continuum (Section 5.3), set theory (Section 5.4), and the natural numbers (Section 5.5), and highlight the familiar view and the challenges posed by nonstandard analysis. As noted at the end of Section 4.4.8, challenging habits of thought may be the cause of some negative attitudes towards nonstandard analysis.

5.1.4 Realism

In mathematical realism, mathematical objects are granted an existence independent of human thought. Depending on the objects to which the realism postulate refers and the way in which existence is understood, there are numerous variants and gradations of mathematical realism.

Metaphysical realism (Platonism)

The classical form of mathematical realism is *metaphysical realism*, also called *mathematical Platonism* in reference to Plato's philosophy. The objects of mathematics, originally natural numbers, geometric objects, and quantities derived from them, therefore belong to an ideal, immaterial, but real world.

In terms of modern mathematics, we encounter this position as *set-theoretical realism*. Cantor, as the founder of set theory, was convinced of the real existence of infinite multiplicities. This deep conviction made him adhere to his new theory of actual infinity despite all opposition (for example, from Kronecker) and despite the examples of incon-

---

[2] J. D. Monk has estimated that 65 % of mathematicians are platonists, 30 % are formalists, and 5 % are intuitionists (and thus also constructivists) ([156], p. 3).

sistent multiplicities (such as the set of all sets or the set of all ordinal numbers) known to him.

Also for Gödel, mathematical objects, such as those of set theory, existed *objectively*. At least Gödel considered the belief in their existence to be just as legitimate as the belief in the existence of physical bodies. In *Russell's mathematical logic*, he said about classes and concepts:

> Classes and concepts may, however, also be conceived as real objects, namely, classes as "pluralities of things" or as structures consisting of a plurality of things and concepts as the properties and relations of things existing independently of our definitions and constructions.
>
> It seems to me that the assumption of such objects is quite as legitimate as the assumption of physical bodies and there is quite as much reason to believe in their existence. They are in the same sense necessary to obtain a satisfactory systems of mathematics as physical bodies are necessary for a satisfactory theory of our sense perceptions and in both cases it is impossible to interpret the propositions one wants to assert about these entities as propositions about the "data", i. e., in the latter case the actually occurring sense perceptions ([81], p. 137).

According to Gödel, we have a kind of "perception" for the objects of set theory, a mathematical intuition through which the axioms of set theory impose themselves on us as *true*.

The philosopher Penelope Maddy initially advocated a set-theoretical realism (based on Gödel) [148], but later revised this position in favor of her *mathematical naturalism* or "thin realism" [147, 149]. According to the latter view, the justification for set-theoretical axioms must come from mathematics itself, not from outside (for example, from philosophy or physics). The justification of the axioms lies in their fruitfulness within mathematics (including set theory itself). At the end of her book *Naturalism in Mathematics*, Maddy writes:

> ...set theory aims to MAXIMIZE and UNIFY because of its foundational role. But it must also include an analysis of the purely set-theoretic goals that motivate the development of set theory on its own terms ...([147], emphasis in the original).

The criterion of fruitfulness brings pragmatic and possibly aesthetic aspects into the evaluation of mathematics. Such a position is not bound to metaphysical realism. The challenge is to justify the extent to which fruitfulness can be assessed objectively.

Empirically grounded realism

While metaphysical realism assumes a source of knowledge independent of experience (otherwise one could not know anything about infinite sets), other forms of realism emphasize a coupling of mathematics to the empirical sciences and thus tie in with John Stuart Mill's empiricist conception [155].

Quine and Putnam see a direct connection between parts of mathematics and physical reality. According to their *indispensability argument*, mathematics is so strongly integrated into physics that it is not possible to be a realist with regard to physical theories without also being a realist with regard to mathematical theories (cf. [176], p. 74). This understanding of realism is less about existence in the metaphysical sense than about the objectivity of knowledge. Putnam describes it as follows:

> The question of realism is the question of the objectivity of mathematics and not the question of the existence of mathematical objects ([176], p. 70).

The existence of objects is a *postulated existence*. For the distinction between *realism in ontology* and *realism in truth-value* see also [195], p. 37.

The German philosopher Torsten Wilholt advocates a partial realism in [227].[3] For him, mathematics is a two-part science with a realistic and a formal, deductivistic part. According to this view, *realistic mathematics* originally arose in unity with its primary applications; its objects are *universals* that have realizations in the world of experience, for example as properties or relations (pp. 282–284). Positive integers are properties of *aggregates* of certain causal processes (pp. 178–193), positive real numbers are properties of real *proportions of magnitudes* (pp. 193–216). For Wilholt, they belong to realistic mathematics.[4] In order to arrive at satisfactory mathematical theories, Wilholt must understand the universals *ante rem*. This means, for example, for the natural numbers: Since every natural number is supposed to have a successor, one has to extrapolate the number properties of physically realized aggregates into the counterfactual. If u is the number property of the maximum possible physical aggregate A, then $u + 1$ is the number property that an aggregate formed from A and P *would* have if there *were* another process P that *could* be added to A (pp. 182–183). In the same way, proportions of magnitudes must be extrapolated into the counterfactual in order to guarantee certain closure properties of the rational and real numbers (p. 207). Analogous to the highly formalizable realistic mathematics, according to Wilholt formal systems can be studied that have no direct reference to reality. Instead of truth, Wilholt speaks here of *acceptability* of the axioms (and their implications) (p. 284).

The actually infinite universe

In this chapter, we understand *realism* to summarize any position that assumes the objective existence of an actually infinite universe of mathematical objects with objective relationships between the objects. Whether existence is understood in a metaphysical sense (Platonism) or in the sense of an objectively true statement of existence without ontological claim (realism in truth-value), whether as (empirically grounded) postulated existence (Quine/Putnam) or as the existence of universals ante rem (Wilholt), should not be decisive for our use of the term *universe*.

---

[3] Wilholt himself calls it "cautious mathematical realism" (in German "behutsamer mathematischer Realismus"). The further page references in this section refer to [227].

[4] Why this realist justification of real numbers is problematic is discussed in [26] (pp. 237–238). We address a further problem in relation to positive integers in Section 5.5.3.

Such a universe can be described axiomatically using a suitable formal language, but because of Gödel's incompleteness theorems, there are always statements that can neither be proved nor disproved.

One consequence of the realist position is that statements about the universe (according to classical logic) are either true or false in an absolute sense, even if we do not know the truth value or if we cannot determine it on the basis of the established axioms for principal reasons (as in the case of undecidable statements). In particular, also statements that quantify over the entire universe are either true or false. Realistic mathematics is *factual mathematics*.

5.1.5 Constructivism

The term *constructivism* is used to summarize different currents that emerged as a reaction to the infinitism of Cantor's set theory and the inconsistencies that occurred therein. These currents include intuitionism, predicativism, finitism, and ultrafinitism (see, for example, [26], pp. 107–109). The unifying element is the requirement that mathematics should be *constructive* and *effective* in a certain sense. Proofs of existence are *constructions*.[5] This rules out abstract set-theoretical concepts that use the axiom of power set or the axiom of choice (see also Section 5.4.10). The infinite is accepted at most in its countable form (predicativism) or in its potential form (finitism) (see Section 5.2.1). Ultrafinitism is even more restrictive.

Intuitionism, founded by Brouwer, rejects classical logic and equates truth with the availability of a (constructive) proof. Indirect proofs or the theorem of the excluded middle are not accepted. "$A \vee B$ is true" means "we have a proof for A or we have a proof for B". This means that $A \vee \neg A$ is not automatically true. Many rules of classical logic do not apply in intuitionistic mathematics. A complete formalization of intuitionistic logic comes from Heyting [93].

Due to the restrictions demanded by constructivism, almost the entire set theory and thus also parts of classical analysis no longer apply. Even the definition of real numbers (see Section 1.3) is not possible as usual. Constructivist versions of calculus exist, for example, by Lorenzen [142] and Bishop [37].

Finitism and primitive recursive arithmetic

Finitism is also relevant in connection with formalism to be discussed in Section 5.1.6, as Hilbert demanded finite methods for *metamathematics* (i.e., the analysis of the foundations of mathematics using mathematical methods). Tait proposed in [213] that finitistic reasoning should be understood as primitive recursive reasoning in the sense of

---

[5] There are various positions on the ontology of these constructions: objectivism, intentionalism, mentalism, nominalism (see [26], p. 109.)

[202] by Skolem. This is also called *primitive recursive arithmetic* (short: PRA). The title of Skolem's work, "The foundations of elementary arithmetic established by means of the recursive mode of thought without the use of apparent variables ranging over infinite domains",[6] also outlines the program. PRA allows the symbolization of arbitrary primitive recursive functions and the formation of quantifier-free formulas. The following induction rule is available as a means of proof for generalizations: From $\varphi(0)$ and $\varphi(x) \Rightarrow \varphi(\sigma(x))$, it is possible to infer $\varphi(y)$, where φ is a quantifier-free formula and σ is the successor function.[7]

The constructivist justification for the induction rule based on potential infinity is that every constructible number (and thus, in the sense of constructivism, every existing number), say $g(0)$ (with a primitive recursive function g), can be resolved to $\sigma \cdots \sigma(0)$ (with finitely many applications of σ). Thus, a proof for $\varphi(g(0))$ consists of a finite number of applications of the induction step $\varphi(x) \Rightarrow \varphi(\sigma(x))$.

For example, to prove a classical arithmetic formula of the form $\forall x \exists y\, \varphi(x, y)$, you have to construct a suitable primitive recursive function f and prove the formula $\varphi(x, f(x))$ by induction.

It should be mentioned at this point that the confidence that every primitive recursion terminates and is thus finitistically unproblematic also means a certain idealization. Nelson (who in this respect belongs to ultrafinitism) has questioned the justification of this confidence (see [159]). In his book *Predicative Arithmetic*, he wrote:

> It appears to be universally taken for granted by mathematicians, whatever their views on foundational questions may be, that the impredicativity inherent in the induction principle is harmless—that there is a concept of number given in advance of all mathematical constructions, that discourse within the domain of numbers is meaningful. But numbers are symbolic constructions; a construction does not exist until it is made; when something new is made, it is something new and not a selection from a preexisting collection. There is no map of the world because the world is coming into being ([161], p. 2).

5.1.6 Formalism

Mathematical formalism, which goes back to Hilbert, distinguishes between finitistic and infinitistic mathematics. Finitistic mathematics is considered certain per se, while

---

[6] In the original German: "Begründung der elementaren Arithmetik durch die rekurrierende Denkweise ohne Anwendung scheinbarer Veränderlichen mit unendlichem Ausdehnungsbereich".

[7] Simpson gives a formal axiom system of PRA with quantifiers in [200], but points out that there is a quantifier-free axiomatization. The induction schema can initially be formulated with bounded quantifiers of the form $\forall x < y$ as follows:
$$\varphi(0) \land \forall x < y \left(\varphi(x) \Rightarrow \varphi(\sigma(x)) \right) \Rightarrow \varphi(y).$$
A formula with only bounded quantifiers is equivalent to a quantifier-free formula in PRA. A separate inference rule for induction is then not necessary.

infinitistic mathematics (especially Cantor's set theory) is considered uncertain in principle and is to be justified by formalization and metamathematical considerations with "finite methods" (Hilbert's program).

Hilbert did not define exactly what he meant by "finite methods". In today's proof theory, the formal system PRA of primitive recursive arithmetic is often used, as it can be understood as a theory of the potentially infinite and provides the essential means required in metamathematics.

Due to Gödel's incompleteness theorems, Hilbert's program could not be realized in its original form. However, certain parts of mathematics can be finitistically justified using the results of reverse mathematics. This can be seen as a *partial realization of Hilbert's program* (see Section 5.1.7).

For a formalist, the actual infinite is a useful fiction. In his article "On the Infinite", Hilbert wrote:

> Our principal result is that the infinite is nowhere to be found in reality. It neither exists in nature nor provides a legitimate basis for rational thought – a remarkable harmony between being and thought ([95], p. 190, as translated in [94]).

Robinson expressed himself in a similar vein:

> My position concerning the foundations of Mathematics is based on the following two main points or principles:
> (i) Infinite totalities do not exist in any sense of the word (i. e., either really or ideally). More precisely, any mention, or purported mention, of infinite totalities is, literally, meaningless.
> (ii) Nevertheless, we should continue the business of Mathematics "as usual," i. e., we should act as if infinite totalities really existed.
> ([180], p. 230.)

According to Robinson, infinite totalities have no *referents* (neither in reality nor in a platonistic ideal world) and their mention is in this sense *meaningless*.[8] In addition to the ontological aspect, the above Hilbert quote also has an epistemological aspect ("It [the infinite] neither exists in nature nor provides a legitimate basis for rational thought").

In this chapter, we understand formalism to be a position that is finitistic in relation to metamathematics and fictionalistic in relation to infinitistic mathematics. The objects of metamathematics (e. g., character strings, metanumbers) are understood as mental constructions whose domain is open and only potentially infinite. No infinite "universe of metamathematics" is assumed as a completed totality. Actually infinite entities of infinitistic mathematics are only granted a *theoretical existence*. Such entities (sets or classes) "exist" only in a formal sense, that is, by convention, within the framework of a formalized axiomatic theory (hoping that the theory is consistent). Nevertheless,

---

[8] In 1975, Robinson clarified his point (i) by stating "that mathematical theories that, allegedly, deal with infinite totalities have no detailed meaning, i. e., reference" ([179], p. 42, reprinted in [184], p. 557).

we talk about infinite totalities within the framework of the axiomatic theory, *as if* they really existed (as per Robinson's recommendation in point (ii) above). Thus, we also use classical logic for quantifications over the entire universe. For example, we know that the continuum hypothesis CH is not decidable in ZFC, but from ZFC follows (according to classical logic) CH ∨ ¬CH.

Just as we have described realistic mathematics as factual mathematics, we can now say: Formalistic mathematics is *hypothetical mathematics*.

5.1.7 Reverse mathematics and Hilbert's program

Reverse mathematics is a research program on the foundations of mathematics that goes back to Harvey Friedman [77] and investigates which mathematical theorems can be proven in which formal systems. Major contributions to this program come from Stephen G. Simpson (see [200]). The philosophical significance of reverse mathematics lies in the fact that it can be used to realize Hilbert's program, at least in part (see [158]).

Simpson distinguishes between *set-theoretic mathematics* and *non-set-theoretic mathematics*. He also calls the latter *ordinary mathematics*. Set-theoretic mathematics refers to the branches of mathematics that only became possible through the fundamental revolution associated with set theory, such as general topology, abstract functional analysis and abstract set theory itself. Ordinary mathematics consists of those branches that are independent of abstract set theory and that had already been studied to a large extent before set theory became the foundation of mathematics. Among these are geometry, number theory, differential equations, real and complex analysis, countable algebra, topology of complete separable metric spaces,[9] mathematical logic, and computability theory. According to Simpson, the distinction between ordinary and set-theoretic mathematics corresponds roughly to the distinction between "countable mathematics" and "uncountable mathematics", if one adds the study of (possibly uncountable) complete separable metric spaces to countable mathematics (cf. [200], p. 1). The following presentation is based on [200].

The system Z_2 of second-order arithmetic

The investigation takes place in second-order arithmetic (abbreviated as Z_2). This system is weaker than fully-fledged set theories such as ZFC or NBG, but is sufficient to reproduce essential parts of classical mathematics. It is therefore well suited for foundational investigations. All mathematical concepts (numbers, functions, etc.) are to be encoded as natural numbers or as sets of natural numbers.

The language L_2 of Z_2 contains two types of variables, *number variables* (denoted by lowercase letters such as i, j, k, m, n) and *set variables* (denoted by uppercase letters such

---

[9] A metric space is called separable if it has a countable dense subset.

as X, Y, Z). In contrast to the language L_1 of first-order arithmetic (Peano arithmetic), L_2 also contains formulas of the form $t \in X$, where t is a numerical term and X is a set variable. Likewise, L_2 allows quantifications over set variables. Quantifiers followed by a number variable (such as $\forall n$, $\exists n$) are called *number quantifiers* and quantifiers followed by a set variable (such as $\forall X$, $\exists X$) are called *set quantifiers*.

In addition to the so-called *basic axioms* (that can be formulated without set quantifiers), Z_2 contains the *induction axiom*

$$\forall X\,((0 \in X \wedge \forall n\,(n \in X \to n+1 \in X)) \to \forall n\,(n \in X)) \tag{5.1}$$

and the *comprehensions schema*

$$\exists X \forall n\,(n \in X \leftrightarrow \varphi(n)), \tag{5.2}$$

where $\varphi(n)$ is any formula of L_2 in which X does not occur freely.

From (5.1) and (5.2) follows the full L_2-induction schema

$$(\varphi(0) \wedge \forall n\,(\varphi(n) \to \varphi(n+1))) \to \forall n\,\varphi(n), \tag{5.3}$$

where $\varphi(n)$ is any formula of L_2.

Important subsystems of Z_2

Five subsystems of Z_2 (Simpson calls them "the Big Five") play a prominent role in reverse mathematics. Sorted in ascending order of strength of proof, these are RCA_0, WKL_0, ACA_0, ATR_0, and Π_1^1-CA_0. They differ in the strength of their respective comprehension schema, i. e., in which sets can be formed. The subscript 0 indicates in each case that the induction schema is restricted compared to (5.3). We will go into the first three subsystems in more detail.

RCA stands for "Recursive Comprehension Axiom". In addition to the basic axioms, RCA_0 contains the Σ_1^0-induction schema (that is, (5.1) for all Σ_1^0-formulas φ)[10] and the following so-called Δ_1^0-comprehension schema:

$$\forall n\,(\varphi(n) \leftrightarrow \psi(n)) \to \exists X \forall n\,(n \in X \leftrightarrow \varphi(n)), \tag{5.4}$$

where $\varphi(n)$ is a Σ_1^0-formula and $\psi(n)$ is a Π_1^0-formula and X does not occur freely in $\varphi(n)$. The name "recursive comprehension axiom" is related to the fact that the smallest ω-model of RCA_0 contains just the recursive sets of $\mathcal{P}(\omega)$, where ω is the set of natural numbers of background set theory and an ω-model is a model with domain ω (cf. [200], p. 65).

---

[10] A Σ_1^0-formula is a formula of the form $\exists m\,\varphi(m)$, a Π_1^0-formula is a formula of the form $\forall m\,\varphi(m)$, where $\varphi(m)$ is an L_2-formula in which all quantifiers are bounded number quantifiers (i. e., of the form $\forall m \le m_0$ resp. $\exists m \le m_0$).

In RCA_0, the number system can already be built up to the real and complex numbers, the intermediate value theorem for continuous functions and Peano's theorem for ordinary differential equations can be proven. The syntax of first-order logic, as well as countable models, can be defined in RCA_0 (by Gödel numbering) and the correctness theorem can be proved (a set of theorems that has a countable model is consistent) (cf. [200], pp. 73–96).

The system WKL_0 contains, in addition to the axioms of RCA_0, an axiom called *Weak König's Lemma*. It states that every infinite binary tree has an infinite path. More precisely, WKL_0 contains an encoding of this statement in the language L_2.

Within RCA_0, one can prove that WKL_0 is equivalent to each of the following (and many other) theorems (cf. [200], pp. 36–37):
- Every continuous real-valued function on $[0,1]$ is uniformly continuous.
- Every continuous real-valued function on $[0,1]$ is Riemann integrable.
- The maximum principle: Every continuous real-valued function on $[0,1]$ attains its maximum.
- Gödel's completeness theorem: Every finite, or countable, set of sentences in the predicate calculus has a countable model.

The abbreviation ACA stands for "Arithmetic Comprehension Axiom". The system ACA_0 contains, in addition to the RCA_0 axioms, the induction schema (5.3) for all arithmetic formulas, i.e., an induction schema like the Peano arithmetic PA (cf. (5.13) in Section 5.4.7). Within RCA_0, one can prove that ACA_0 is equivalent to each of the following (and many other) theorems (cf. [200], pp. 34–35):
- Every bounded sequence of real numbers has a least upper bound.
- The Bolzano–Weierstrass theorem: Every bounded sequence of real numbers (or of points in $\mathbb{R}^n$) has a convergent subsequence.

A partial realization of Hilbert's program

As mentioned in Section 5.1.6, certain parts of mathematics can be finitistically justified with the results of reverse mathematics, which can be seen as a partial realization of Hilbert's program. We summarize the most important statements from [200], pp. 369–379.

Definition 25. A formal system S is called *finitistically reducible* if all Π_2^0-sentences that are provable in S are also provable in PRA.

A Π_2^0-sentence is a sentence of the form $\forall m \exists n\, \varphi(m,n)$, where $\varphi(m,n)$ is an L_2-formula in which all quantifiers are bounded number quantifiers.[11] Among the "Big Five", exactly RCA_0 and WKL_0 are finitistically reducible.

---

11 For the definition of the arithmetical hierarchy, see, e.g., [197], Chapter 7.5.

The finitistic reducibility of WKL_0 means: Every Π_2^0-sentence provable in WKL_0 is already provable in PRA or, in other words, WKL_0 is conservative over PRA with respect to Π_2^0-sentences.[12] Furthermore, the conservativeness of WKL_0 over PRA is provable (encoded as a Π_2^0-sentence) in WKL_0 and thus in PRA. If finite methods are identified with PRA, this result of reverse mathematics can therefore be understood as a partial realization of Hilbert's program.

5.1.8 Mathematical practice

Does a mathematician have to choose a philosophical position? Or can he take an agnostic or even an ambivalent or oscillating position?

According to Davis and Hersh, the prevailing opinion is that the typical mathematician "is a Platonist on weekdays and a formalist on Sundays" ([58], p. 359). In other words, as long as he works mathematically, he is convinced that he is exploring an objective reality; if he is to explain this reality philosophically, he prefers to pretend that he does not ultimately believe in such a reality. Paul Cohen writes:

> The Realist position is probably the one which most mathematicians would prefer to take. It is not until he becomes aware of some of the difficulties in the set theory that he would even begin to question it. If these difficulties particularly upset him, he will rush to the shelter of Formalism, while his normal position will be somewhere between the two, trying to enjoy the best of two worlds ([49], p. 11).

According to Shapiro, most mathematicians are not the least bit interested in philosophy. He calls it the *philosophy-last-if-at-all principle* ([195], p. 7). He characterizes the working mathematician as a "working realist" (analogous to the working-day Platonist in Davis and Hersh):

> I define a *working realist* to be someone who uses or accepts the inferences and assertions suggested by traditional realism, items like excluded middle, the axiom of choice, impredicative definition, and general extensionality (ibid.).

Realists and formalists therefore agree on the "working level". As long as what is being investigated can be modeled in ZFC, it is "real" or can be treated as such (constructivists do not agree, but are probably in the minority). Simpson describes the role of ZFC as follows:

> The ZFC formalism provides two extremely important benefits for mathematics as a whole: a common framework, and a common standard of rigor ([201], p. 6).

---

[12] If there is a proof of the Π_2^0-theorem $\forall m \exists n\, \varphi(m, n)$ in WKL_0, then there is a proof of $\varphi(m, f(m))$ with a primitive recursive function f in PRA (cf. [198], p. 8).

The "working realist" in us shapes our thinking. This also influences our view of nonstandard analysis, because nonstandard theories and especially nonstandard extensions of ZFC seem to call our realism into question. We will examine this in more detail in the next sections.

5.2 Infinity

5.2.1 Potential vs. actual infinity

Colloquially, we often use the term "infinite" in the sense of "very much" or "very large", for example, when we invest an infinite amount of effort in something or are infinitely grateful to someone. Even the number of grains of sand on the beach sometimes symbolizes infinity. But Archimedes already calculated how many grains of sand would be needed to fill the entire cosmos (which was according to the ideas of the time limited by the celestial sphere). The number was huge compared to the numbers used until then and required a new way of representation, but it was not infinite.

The distinction between potential and actual infinity goes back to Aristotle. He only allowed for *potential infinity* as something that can be increased (or decreased) at will, but not an *actual* or *completed infinity* that is given as a whole.

Using the example of natural numbers: The counting sequence 1, 2, 3, ... can be continued indefinitely, there is no last number that would not allow any further additions. But an infinite counting sequence as a completed whole does not exist. Or using a geometric example: The straight line defined by two points can be extended to both sides as desired, but an infinite straight line as a whole does not exist.

Today, we can hardly comprehend the fear of actual infinity—at least if we are mathematicians. Infinite sequences or infinite lines, infinite sets in general, have become self-evident objects of mathematical work. Viewed impartially, however, the Aristotelian position is still very plausible. How could something that is without end ever be completed? It is and remains a paradox—a paradox to which we have become accustomed and which we even explicitly postulate in ZFC with the axiom of infinity.

Since the triumph of Cantor's set theory, potential infinity seems to have been relegated to the role of a historical concept. However, this does not apply to the mathematical foundations and the philosophy of mathematics. Stephen G. Simpson compares four philosophical positions with regard to their attitude towards potential or actual infinity.

Ultrafinitism: Infinities, both potential and actual, do not exist and are not acceptable in mathematics.

Finitism: Potential infinities exist and are acceptable in mathematics. Actual infinities do not exist and we must limit or eliminate their role in mathematics.

Predicativism: We may accept the natural numbers, but not the real numbers, as a completed infinite totality. Quantification over $\mathbb{N}$ is acceptable, but quantification over $\mathbb{R}$, or over the set of all subsets of $\mathbb{N}$, is unacceptable.

Infinitism: Actual infinities of all kinds are welcome in mathematics, so long as they are consistent and intuitively natural.
([198], p. 3.)

Simpson himself advocates finitism, as he considers it most compatible with his philosophical position of objectivism, which he describes as follows: The epistemology of objectivism calls for a close relationship between *existence* and *consciousness*, between a reality that exists (independently of our consciousness) and our conscious act of will in grasping objects. According to Simpson, mathematics should strive for an objective understanding of the mathematical aspects of reality. It is desirable to trace mathematics—or at least the applicable parts of mathematics—back to an objective basis (cf. [201]). He pursues this goal within the framework of reverse mathematics (see Section 5.1.7).

We will return to the significance of potential infinity in Section 5.5.

5.2.2 Arithmetic vs. unarithmetic infinity

The idea of actual infinity has sometimes been associated—also by mathematicians—with philosophy and theology and was then distinguished from actual infinity in mathematics, for example in Cantor (cf. [44], p. 378). In his mature system, Leibniz distinguished between mathematical extension, which is *potentially* infinitely divisible, and the matter of physical bodies, which is *actually* infinitely divided (cf. [116], p. 40).

We focus here on actual infinity in mathematics, and specifically on an aspect that is important for the distinction between standard and nonstandard analysis. It is about the Euclidean axiom "The whole is greater than its part" (hereafter *part-whole axiom*).

Counting to infinity

Is it possible to think of counting to infinity and beyond? Cantor is known to have done this, both in the ordinal and cardinal sense, and thus arrived at his transfinite ordinal and cardinal numbers.[13] In Zermelo–Fraenkel's set theory ZF, the natural numbers are defined set-theoretically according to von Neumann by $0 := \emptyset$ and $z+1 := z \cup \{z\}$. The set ω of all natural numbers is then defined as the smallest set that contains 0 and for each element z also its successor $z + 1$. More precisely, one defines:

$$x \text{ is inductive} \quad :\Leftrightarrow \quad (\emptyset \in x \land \forall z\, (z \in x \Rightarrow z \cup \{z\} \in x)),\text{[14]}$$

---

[13] The following explanations of ordinal and cardinal numbers can be found in textbooks on set theory, for example, in [106], Chapters 2 and 3.

[14] This is the usual definition in ZF. In Section 1.3.2 we had defined *inductive subsets of* $\mathbb{R}$ in such a way that they contain 1 and with each number also its successor. Therefore, the context may need to be considered.

and then

$$\omega := \{z \mid \forall y \, (y \text{ is inductive} \Rightarrow z \in y)\}.$$

According to the axiom of infinity, an inductive set exists. Therefore, ω can be obtained by separation. The set ω contains just those elements that are contained in *every* inductive set. In other words, for each inductive set x, we have $\omega \subseteq x$. Since the set ω itself is inductive, it is the *smallest inductive set* (in the sense of set inclusion).

The relation $\in_\omega$ induced by the predicate $\in$ on ω corresponds to the less-than relation on the natural numbers, because by construction each $z \in \omega$ is the set of its $\in_\omega$-predecessors. The relation $\in_\omega$ is even a *well-order relation* on ω. This means that $\in_\omega$ is a total order relation on ω and each nonempty subset of ω contains a least element (with respect to $\in_\omega$).

Also ω itself is the set of its $\in_\omega$ predecessors, but has no immediate predecessor. It is therefore obvious to continue the counting process *after* all elements of ω with the set ω and then again according to the principle $z + 1 := z \cup \{z\}$. This leads to *transfinite ordinals*.

In general, x is an *ordinal number* (short: Oz x) if $\in_x$ is a well-order relation on x and each element of x is the set of its $\in_x$-predecessors. An ordinal number x is called *successor ordinal* if it has an immediate predecessor, i.e., if there is an ordinal number z with $x = z \cup \{z\}$. It is called *limit ordinal* if it is nonzero and not a successor ordinal.

It can be shown in ZF that for every ordinal number there is an ordinal number with larger cardinality and that the ordinal numbers form a proper class. The essential order properties of ordinal numbers can be transferred to the entire class (with $\in$ as the order predicate). This is usually written $<$ instead of $\in$. Two ordinal numbers α, β are always comparable, i.e., either $\alpha < \beta$ or $\alpha = \beta$, or $\beta < \alpha$. The elements of ω are by definition the *finite* ordinal numbers. All other ordinal numbers are called *infinite*. This means that ω is the smallest infinite ordinal number and also the smallest limit ordinal.

Cardinal numbers are those ordinal numbers for which the cardinality jumps to the next higher level. They are defined by the *aleph operation*.[15] The arguments are usually written here as subscripts. One defines $\aleph_0 := \omega$ and $\aleph_{\alpha+1}$ as the smallest ordinal number that has larger cardinality than $\aleph_\alpha$. For limit ordinals δ, one defines $\aleph_\delta := \bigcup \{\aleph_\beta \mid \beta < \delta\}$. To ensure that the operation is defined for the entire universe, one defines $\aleph_x := \emptyset$ if x is not an ordinal number. A set x is a *cardinal number* if $x \in \omega$ or if there is an ordinal number α with $x = \aleph_\alpha$. In the former case, x is called a *finite cardinal*, in the latter an *infinite cardinal*.

In ZFC it can be shown (and the axiom of choice is essential) that every set x is equipotent to a unique cardinal number. This cardinal number is denoted by $|x|$

---

15 The global recursion theorem for ordinal numbers (see [106], p. 22) enables recursive definitions of operations on the set universe.

or card(x) and is called the *cardinality* of x. In this sense, cardinal numbers are a measure of the "size" of infinite sets. They "count" the elements of infinite multiplicities.

You can calculate with cardinal numbers in a certain way, although not in the usual way. Cardinal addition, multiplication, and exponentiation are defined as follows:

$$\kappa + \mu := |(\kappa \times \{0\}) \cup (\mu \times \{1\})|,$$
$$\kappa \cdot \mu := |\kappa \times \mu|,$$
$$\kappa^\mu := |\kappa^\mu|.$$

The cardinal operations extend the ordinary arithmetic operations defined on ω. The addition and multiplication of cardinals remains commutative and associative even in the transfinite, but strangely "unarithmetic" rules apply. For cardinals κ, μ, at least one of which is transfinite, $\kappa + \mu = \max\{\kappa, \mu\}$ and (if neither is 0) also $\kappa \cdot \mu = \max\{\kappa, \mu\}$. Roughly speaking, the rules reflect the fact that the formation of unions or Cartesian products in the transfinite does not increase cardinality (unlike what is known from the finite). The part–whole axiom is violated. From the definition of cardinal exponentiation, it follows that the power set of an infinite set A has cardinality $2^{|A|}$ (as known from finite sets). This is always larger than $|A|$.

Cantor's transfinite ordinal and cardinal numbers are standard in mathematics today. Is there another way to count to infinity and beyond? Leibniz and Johann Bernoulli already discussed in an exchange of letters whether an infinite series must have an infinitieth term (and then further terms). While Bernoulli was clearly in favor of this, Leibniz was reluctant and argued that this was at least not a compelling conclusion.[16] Bernoulli's infinitieth terms are in an ordinal sense at an infinite rank. Nevertheless, there is an essential difference to Cantor's transfinite ordinal numbers. While the former always have successors and predecessors, this does not apply to transfinite ordinals. They have successors, but no predecessors if they are limit ordinals. So here we have two completely different concepts of counting to infinity. Infinite totalities, and thus the possibility of infinite cardinal numbers, were generally rejected before Cantor, as they violate the part–whole axiom (as Galileo already stated).

Measuring the infinite

Is it possible to extend the idea of measurement to infinity and beyond? Are there infinite quantities? Cantor's cardinal arithmetic does not provide an answer to this question, because the multiplication of a transfinite cardinal number by a quantity (or a real number) is not defined there.

---

[16] Spalt discusses the correspondence in [206] (pp. 51–58) and comes to the conclusion that Leibniz rejected an infinite number. However, Leibniz did not claim to Bernoulli that there could not be an infinitieth term, but only that such a term did not necessarily have to exist.

Leibniz distinguished between an *unbounded infinite line* (linea infinita interminata) and a *bounded infinite line* (linea infinita terminata). According to Leibniz, no quantity (i. e., no length) can be attributed to an unbounded infinite line because such a line would have to be as large as a part of itself. For example, if you divide a straight line at a point and move the half-straights a little apart, you get a straight line that is missing a piece. However, dividing and moving would not change the quantity. As with infinite totalities, the part–whole axiom would be violated.

A bounded infinite line can be imagined as a line segment from one point to a (fictitious) infinitely distant point. In the case of a shift, the infinitely distant end point would also be shifted, and no proper part of the original line would be created. Leibniz allowed such bounded infinite lines as objects of geometry. He described their lengths—infinite quantities—as useful fictions and used them in his infinitesimal calculus (see [138], p. 61). Calculating with these fictitious quantities works in the same way as calculating with ordinary quantities. The reciprocals of such infinite quantities are infinitesimal.

A completely different handling of infinite quantities is common in today's measure theory (see, for example, [187], pp. 9–10). There, one defines the *extended real number system* $\overline{\mathbb{R}} := \mathbb{R} \cup \{-\infty, +\infty\}$ and $-\infty < x < +\infty$ for all $x \in \mathbb{R}$. The following rules apply when calculating with the infinite elements:

$$x + (\pm\infty) = (\pm\infty) + x = \pm\infty, \quad \text{for } x \in \mathbb{R},$$
$$(\pm\infty) + (\pm\infty) = \pm\infty,$$
$$x \cdot (\pm\infty) = (\pm\infty) \cdot x = \pm\infty, \quad \text{for } 0 < x \leq +\infty,$$
$$x \cdot (\pm\infty) = (\pm\infty) \cdot x = \mp\infty, \quad \text{for } -\infty \leq x < 0,$$
$$0 \cdot (\pm\infty) = (\pm\infty) \cdot 0 = 0.$$

Note that $(\pm\infty) + \mp\infty$ is not defined. Division by $\pm\infty$ is also not defined. The Lebesgue measure on $\mathbb{R}$ can take values from $[0, +\infty]$. The unbounded infinite lines rejected by Leibniz have the measure $+\infty$ here.

The part and the whole

In his book "Analysis im Wandel und Widerstreit" (Analysis in Transition and Dispute), Detlef D. Spalt states:

> What characteristic(s) does the "actual" infinity have? Fatally, this is completely unclear! ([207], p. 676.)[17]

According to Spalt, there are (at least) two fundamentally different and mutually exclusive views on this, one that retains the Euclidean axiom "The whole is greater than its part" and one that rejects it.

---

17 In the original German: "Welche kennzeichnende(n) Eigenschaft(en) hat das „aktuale" Unendlich? Fatalerweise ist dies völlig unklar!"

The first view allows an arithmetic infinity, as used in nonstandard analysis. Then, for example, the set of integers from 1 to an even infinite number v contains twice as many integers as even numbers. The distance from 0 to an infinitely distant point μ is twice as long as the distance from 0 to $\frac{\mu}{2}$, and the distance from 1 to μ is 1 shorter than the distance from 0 to μ.

The second view allows mathematical objects that are as large as a proper part of themselves. The violation of the Euclidean axiom becomes a characteristic property of the actual infinite in set theory (a set is called Dedekind-infinite if it can be mapped bijectively to a proper subset). The infinite set of natural numbers contains as many even numbers as there are numbers in total.

The result is an unarithmetic infinity, which cannot be used for calculations as usual. Transfinite cardinal numbers have no additive and no multiplicative inverse. Also $\mu + \lambda = \kappa + \lambda$ does not imply $\mu = \kappa$. And $\mu \cdot \lambda = \kappa \cdot \lambda$ with $\lambda \neq 0$ does not imply $\mu = \kappa$. The same restrictions apply to $\pm\infty$ in the extended real number system.

Exclusion or coexistence

With the implementation of Cantor's set theory, mathematics decided to accept the actual infinite and to abandon the part–whole axiom. In the arithmetic of the continuum, the decision was made to abandon the actual infinite and to retain the Archimedean property in the definition of the real numbers: There are arbitrarily large, but no infinitely large real numbers (meaning here: no real numbers larger than all natural numbers). Although non-Archimedean fields continued to be studied (see [69]), they no longer played a major role in the development of calculus.

However, the decision in favor of one option does not exclude the other, as nonstandard analysis proves. In contrast to Spalt's claim, the two conceptions of the actual infinite do not appear to be mutually exclusive alternatives, but rather merely two different but coexisting concepts of infinity. A set A has a certain cardinality $|A|$ and—if A is hyperfinite—also a certain hypernatural number $^{\#}A$ of elements (cf. Definition 16 in Section 3.4.7). Likewise, both concepts of infinity coexist in measure theory. An unbounded interval has the measure $+\infty$. The bounded interval $[0, \mu]$ (with $\mu \gg 1$) has the infinite hyperreal length μ. In internal set theory, arithmetic infinity does not occur at all, but becomes a subform of *finite*. Finite sets can have an unlimited number of elements. Finite real intervals can have an unlimited length.

5.2.3 Challenge: The actual infinite in arithmetic

The familiar view
- Infinite sets are self-evident objects of mathematics.
- Although infinitely large and infinitely small numbers are possible in non-Archimedean field extensions of $\mathbb{Q}$ or $\mathbb{R}$, they hardly play a role in practice.

Challenges posed by nonstandard analysis
- Infinitely large and infinitely small numbers are self-evident objects of mathematics.
- The arithmetic infinite (which obeys the part–whole axiom) and the cardinal infinite (which violates the part–whole axiom) stand side by side on an equal footing and are equally useful concepts in mathematics.

5.3 The continuum

5.3.1 The essence of the continuum

In everyday life, we experience continuity as the flow of time or as the extension of the space surrounding us. In geometry, continua are considered to be lines, surfaces or solids. They are idealizations of everyday continua such as physical lines, surfaces and solids. We use the term "the continuum" here as an abstract generic term for something continuous. The question of what the continuum *is* in its essence has been answered very differently throughout history.[18]

The different views can be divided into atomistic and nonatomistic. According to the atomistic views, the continuum is composed of atoms, i. e., of elementary components that cannot be further divided. Atoms are primitive, the continuum is something composite (i. e., not primitive). Depending on whether the continuum is considered to be only finitely divisible (Democritus) or infinitely divisible, the atoms are of finite quantity or infinitely small.

A special form of the atomistic view is the uncountable atomism that is common today.[19] It was only made possible by Cantor's set theory with its cardinalities beyond the countable. According to this view, the atoms are points without extension. This means that points (as components of a line) have the length 0. This creates firstly the paradox that a multiplication of 0 must result in something positive and secondly the paradox that this positive is not uniquely determined. With a naive view (and $\mathfrak{c}$ as the cardinality of the continuum), for example, $\mathfrak{c} \cdot 0 = 1$ for the line segment from 0 to 1 and $\mathfrak{c} \cdot 0 = 2$ for the equipotent line segment from 0 to 2. Of course, we cannot calculate with cardinal numbers and lengths in this way (according to cardinal arithmetic, $\mathfrak{c} \cdot 0 = 0$ and $\mathfrak{c} \cdot 1 = \mathfrak{c} \cdot 2 = \mathfrak{c}$), but this only shows that we cannot intuitively grasp an uncountable set of points (in contrast to a line segment).

Note that even according to the atomistic view of the continuum, *composition* is more than mere *collection*. Therefore, for example, the *set* $\mathbb{R}$ must be given its structural properties (order, field properties, metric) separately in the form of relations (i. e., further sets).

---

18 For an overview, see, for example, [26], Chapter 3.3.
19 Bedürftig and Murawski call it *trans-transfinite atomism* ([26], p. 194).

According to the nonatomistic view, the continuum is not composed of indivisible points or atoms, but is itself primary, an object of its own kind. This is the prevailing view from the time of Aristotle until the nineteenth century. According to Aristotle, a continuum is always divisible, and only into continua. For him, it is therefore impossible for something continuous to consist of something indivisible. Points as indivisible noncontinua are therefore not already *present* in a continuum (as its components), but they are *put* into the continuum. They divide or limit linear continua, but they do not constitute them.

5.3.2 Continuum vs. set

When we talk about *the* continuum in mathematics today, we usually mean the set $\mathbb{R}$ of real numbers, but this is a relatively recent view of the continuum, which only emerged at the end of the nineteenth century. The prerequisites were the development of set theory by Georg Cantor and the set-theoretical construction of the real numbers, essentially by Cantor and Dedekind.

The original ideas of *continuum* and *set* are actually mutually exclusive. According to Cantor, a set is any collection into a whole of *definite* and *separate* objects of our intuition or our thought. The classical continuum, on the other hand, is something homogeneous in which no definite and separate objects can be identified as components of the continuum. It was not until Cantor, Dedekind, and Hilbert (prepared by Bolzano) that the continuum was declared to be a point set and thus made accessible to set theory. However, this is a decisive change in thinking (see [26], Chapter 3.3).

5.3.3 The Cantor–Dedekind postulate

In today's dominant set-theoretical mathematics, the intuitive, geometric continuum is no longer an object of study. It is replaced by the Dedekind complete ordered field of real numbers. The reference to intuition is established by the concept of the *number line*, which (according to the atomistic concept of the continuum) consists of points without extension, whereby each point corresponds to a unique real number, and vice versa. Only the intuitionists, who reject a set-theoretical reconstruction of the continuum as an uncountable point set, regard the continuum as an independent intuition.

Wilholt, who distinguishes between realistic and formal mathematics in his partial realism, counts the real numbers as realistic mathematics and identifies the positive real numbers (as universals ante rem) with physical proportions of magnitudes (see Section 5.1.4).

A realistic or at least close-to-reality idea of real numbers is probably widespread among mathematicians and users of mathematics. It is apparently justified by the successful use of real numbers in the natural sciences. We can express it succinctly by the following equation:

$$\text{(idealized) physical straight line}$$
$$= \text{geometric straight line}$$
$$= \text{real number line.}$$

Following the article [75], we call this equation the *Cantor–Dedekind postulate*.

5.3.4 The Archimedean axiom

According to Euclid's Definition 4 from Book V of the Elements, two magnitudes have a relationship to each other if they can exceed each other when multiplied ([74], p. 91). In the preceding Definition 3, it is restricted that the magnitudes must be of the same kind, for example, two lengths, two angles, two areas, or two volumes.

Euclid *defines* here what it means that two magnitudes of the same kind have a ratio to each other, he does not claim that two magnitudes of the same kind always have a ratio to each other (i. e., the "Archimedean axiom"). In Book III, Proposition 16, he gives a counterexample by stating that the angle between a circle and a tangent is smaller than any rectilinear angle ([74], p. 57). Robinson wrote the following about the Archimedean axiom in Euclid:

> In fact, Euclid did not accept this axiom at all explicitly but instead introduced a definition (Elements, Book V, Definition 5)[20] which implies that he did not wish to exclude the possibility that magnitudes which are non-archimedean relative to one another actually exist, but that he deliberately confined himself to archimedean systems of magnitudes in order to be able to develop the theory of proportions and, to some extent, the method of exhaustion ([185], p. 191, original spelling preserved).

Leibniz refers to Euclid's Definition 4 in various places, for example, in his definition of *incomparable magnitudes*. In a letter dated June 14/24, 1695, he writes to de l'Hospital:

> I use the term *incomparable magnitudes* to refer to [magnitudes] of which one multiplied by any finite number whatsoever, will be unable to exceed the other, in the same way [adopted by] Euclid in the fifth definition of the fifth book [of *The Elements*] ([133], as translated in [117], p. 262, emphasis in the original).[21]

This, too, is initially only a definition and not an assertion that there are incomparable magnitudes or quantities. The significance of this definition for Leibniz's concept

---

20 Definition 4 in today's editions.
21 In the original French: "J'appelle *grandeurs incomparables*, dont l'une multipliée par quelque nombre fini que ce soit, ne sçauroit exceder l'autre; de la même façon qu'Euclide la pris dans sa cinquieme definition du cinquiéme livre" ([133], pp. 416–417, original spelling preserved, emphasis in the original).

of continuum is the subject of controversial debate among historians. According to the syncategorematic reading, it is an empty definition that does not apply to anything, as statements with infinitesimal or infinite quantities are only to be understood as abbreviations for more complicated statements with ordinary, finite quantities. According to a formalistic reading, the definition describes *unassignable quantities* like the infinitesimal differentials or the lengths of the bounded infinite lines, which can be thought of as fictions in addition to the ordinary quantities (cf. [177], on the one hand, and [16, 17], on the other).

Today, ordered fields are called *Archimedean* if they satisfy the *Archimedean axiom*, i. e., if for all positive field elements x, y there exists an $n \in \mathbb{N}$ such that $nx > y$. In axiomatic introductions to real numbers, the Archimedean axiom is either required separately or inferred from the supremum axiom (i. e., the least-upper-bound property).

Whether the continuum has the Archimedean property (i. e., whether the Archimedean axiom holds) depends on what is meant by the continuum. If it is equated with $\mathbb{R}$ according to the Cantor–Dedekind postulate, then it has the Archimedean property *by definition*. If one sees in $\mathbb{R}$ only a possible (but not mandatory) set-theoretic model of the intuitive continuum, the possibility opens up to consider non-Archimedean models such as $^*\mathbb{R}$ at will.[22]

However, even in the first case, the existence of infinitesimals is not ruled out. This option results from the fact that the set $\mathbb{N}$ of natural numbers is used to formulate the Archimedean property and that there could be unlimited natural numbers. Their reciprocals are infinitesimal. We had already pointed this out at the end of Section 1.3.3.

5.3.5 Challenge: The non-Archimedean continuum

The familiar view
- The geometric straight line consists of uncountably many points, each of which has no extension.
- It has an Archimedean order.
- Once 0 and 1 have been put arbitrarily, the points can be identified with the real numbers (Cantor–Dedekind postulate).

Challenges posed by nonstandard analysis
- The geometric straight line is not a point set. It is *not a set* at all, but a *guiding idea* in its own right, independent of the concept of set.

---

[22] Philip Ehrlich has shown that Conway's system **No** of *surreal numbers* is in some respects the maximum possible model of the continuum: It is the *absolute arithmetic continuum modulo NBG* (von-Neumann–Bernays–Gödel set theory with global axiom of choice). Any ordered field (whose universe is a class in NBG) can be embedded there [68].

- Set-theoretical constructions can emulate certain aspects of this guiding idea and in this sense are *models* of the geometric straight line.
- Standard and nonstandard models are on an equal footing and are equally useful for mathematics.

5.4 Set theory

Set theory is practically "reality" for most mathematicians today, regardless of whether their philosophical position is realistic or formalistic (see Section 5.1.8). Therefore, it deserves special attention when it comes to possible mathematically or philosophically justified reservations against nonstandard extensions of set theory, as presented in the Sections 3.5 and 3.6. The distinction between background set theory and object set theory necessarily plays a role here. Also relevant are substantive ideas such as the cumulative hierarchy, the interpretability of one set theory in another or multiverse theories.

5.4.1 The relevance of set theory in mathematics

In the phase of the establishment of set theory, at the end of the nineteenth and beginning of the twentieth century, it turned out that all mathematical concepts used until then (especially numbers, quantities, curves, functions, and relations) could be defined in terms of set theory—if one was prepared to make certain concessions. In fact, the concessions were enormous, even if they are no longer perceived as such today. Infinite totalities had to be accepted and the part–whole axiom had to be abandoned. The geometric, homogeneous continuum had to be atomized and the geometric quantities had to be reinterpreted as set-theoretically constructed real numbers. On the other hand, the advantages were also enormous. With an axiomatic set theory based on ZFC, "a common framework and a common standard of rigor" was available for mathematics, as we quoted Simpson in Section 5.1.8.

In principle, one can get by with pure set theory without urelements. This gives set theory a special and fundamental role within mathematics. Discussions about the foundations of set theory can thus always be understood as discussions about the foundations of mathematics as a whole.[23]

Another advantage from the point of view of mathematical logic is that set theory can be formalized in a very simple language, a first-order language with $\in$ as the only nonlogical symbol. This means that (apart from the primitive notion *set* itself) every mathematical concept can in principle be formally defined by an $\in$-formula and every mathematical statement can be formalized as an $\in$-sentence.

---

[23] In addition to set theory, there are alternative foundational programs for mathematics, for example, from category theory or type theory (see [11]).

Since functions and relations are sets by definition, we can also quantify over such objects, although only one sort of variable (individual variables) is available. The essential limitation of first-order languages is thus circumvented to a certain extent. At the same time, the advantages that characterize first-order languages in mathematical logic are retained. In particular, important model-theoretical theorems such as the compactness theorem or Gödel's completeness theorem apply.

5.4.2 The set universe as a cumulative hierarchy

In ZF (the axiom of choice is not needed), the axiom of regularity ensures the hierarchical structure of the set universe. If we start with the simplest set, the empty set, continue with successive power set formation and (in the limit case) subsequent formation of the union and extrapolate this procedure into the transfinite, then the axiom of regularity over the remaining axioms (i. e., over ZF^0) is equivalent to the statement that the entire universe is exhausted in this way.

The global recursion theorem for ordinal numbers (see, for example, [106], p. 22) enables recursive definitions of operations on the set universe. An important application of this theorem is the definition of von Neumann's hierarchy V. As with the Aleph operation (see Section 5.2.2), the arguments here are usually written as subscripts.

Definition 26 ([106], p. 64).

$$V_x = \emptyset \quad \text{if } \neg Oz\, x;$$
$$V_0 = \emptyset;$$
$$V_{\alpha+1} = \mathcal{P}(V_\alpha);$$
$$V_\delta = \bigcup\{V_\beta \mid \beta < \delta\} \quad \text{if } \delta \text{ is a limit ordinal.}$$

Definition 27 (ibid.). $Vx :\leftrightarrow \exists \alpha\, x \in V_\alpha$.

By convention, lowercase Greek letters are variables for ordinal numbers. Accordingly, $\exists \alpha$ is an abbreviation for $\exists \alpha\, (Oz\, \alpha \wedge \dots)$. Instead of "$Vx$", the class notation "$x \in V$" is often used. Over ZF^0, the axiom of regularity is equivalent to $\forall x\, x \in V$ ([64], p. 110). In ZF and ZFC, the universe of sets is therefore identical to the class V.

Von Neumann's hierarchy makes the cumulative, hierarchical structure of the set universe particularly clear.[24] The universe rises in levels above the empty set through successive power set formation and (in each limit case) union formation. Each level includes all previous levels.

The fact that the ZF universe is the same as the class V of the cumulative hierarchy is often taken as an indication that the axiomatic theory is based on an intuitively coherent

---

[24] Scott's axiom system captures the idea of cumulative hierarchy axiomatically. This axiom system turns out to be equivalent to ZF (see [64], pp. 166–174).

concept for the notion of set, so that it seems difficult to imagine that ZF or ZFC are contradictory (see, for example, [64], p. 153). However, a *proof* of consistency within the theory itself is not possible according to Gödel's incompleteness theorems.

5.4.3 The constructible hierarchy

In addition to von Neumann's hierarchy V, the *constructible hierarchy* L with the corresponding class **L** of *constructible sets* plays an important role in model-theoretical considerations. Gödel introduced them in 1938 to prove that the axiom of choice AC, the continuum hypothesis CH ($2^{\aleph_0} = \aleph_1$), and the generalized continuum hypothesis GCH ($\forall \alpha\, 2^{\aleph_\alpha} = \aleph_{\alpha+1}$) are consistent relative to ZF (see [82, 80, 83]).

The idea of the constructible hierarchy is that each level $L_{\alpha+1}$ is not the complete power set of L_α, but contains only those sets that can be *defined* by an ϵ-formula based on the level L_α. Since it is not possible to speak directly about formulas in the language of set theory (see Section 5.4.5), the syntax and semantics of formal languages must be reproduced by set-theoretic means, similar to the way in which $0 := \emptyset$ and $z + 1 := z \cup \{z\}$ are used to reproduce the naive natural numbers in set theory. Further details on this and the following explanations can be found in [64], pp. 182–194. In ZF, a unary operation Defpot can thus be defined, which is a "constructible version" of the power set operation $\mathcal{P}$. More precisely, Defpot satisfies the following:

$$\forall y\, \mathrm{Defpot}(y) \subseteq \mathcal{P}(y) \tag{5.5}$$

and for each ϵ-formula $\varphi(z, \overset{n}{x})$,

$$\forall y\, \forall \overset{n}{x}\, (x_1 \in y \wedge \cdots \wedge x_n \in y \rightarrow \{z \in y \mid [\varphi(z, \overset{n}{x})]^y\} \in \mathrm{Defpot}(y)). \tag{5.6}$$

Here, $\overset{n}{x}$ stands for $x_1, \ldots, x_n$ and $[\varphi(z, \overset{n}{x})]^y$ for the ϵ-formula that results from $\varphi(z, \overset{n}{x})$ by relativizing all quantifiers to y (for example, $\forall x_1$ becomes $\forall x_1 \in y$). Analogous to the Definitions 26 and 27, the operation L and the class **L** are then defined as follows:

Definition 28 ([64], p. 184).

$$L_x = \emptyset \quad \text{if } \neg \mathrm{Oz}\, x;$$
$$L_0 = \emptyset;$$
$$L_{\alpha+1} = \mathrm{Defpot}(L_\alpha);$$
$$L_\delta = \bigcup \{L_\beta \mid \beta < \delta\} \quad \text{if } \delta \text{ is a limit ordinal.}$$

Definition 29. $\mathbf{L}x :\leftrightarrow \exists \alpha\, x \in L_\alpha$.

Due to the restriction to constructible sets, the class **L** is much more manageable than the class **V**. For example, it is possible to explicitly define a binary predicate in ZF that well-orders L ([64], pp. 190–192). Furthermore, all axioms of ZF, the axiom of choice and the generalized continuum hypothesis (each relativized to L) hold in L. We also say that L is an *inner model* of ZF+AC+GCH with respect to ZF because every model of ZF includes a model of ZF+AC+GCH (namely the part described by L). This implies: If ZF is consistent, then ZF+AC+GCH is also consistent.

In 1963, Cohen presented his newly developed forcing method and thus showed that the negations ¬AC and ¬CH (i. e., also ¬GCH) are also consistent relative to ZF (see [51, 52, 50]). Together with Gödel's result, this means that AC, CH, and GCH are independent of ZF.

The assumption that the constructible hierarchy already exhausts the entire set universe, i. e., $\forall x\, Lx$, is referred to as the *axiom of constructibility*, or **V** = **L** for short. Since **L** is an inner model of **V** = **L** with respect to ZF, the constructibility axiom is also consistent relative to ZF.

5.4.4 Variety of models

Since the cumulative hierarchy V arises from the empty set through clearly determined, recursively defined steps (formation of power sets or, in the limit case, unions) and in this way exhausts the entire ZF universe, one could get the impression that the universe is thus clearly determined. However, this is not the case. If ZF is consistent, then according to the theorem of Löwenheim, Skolem, and Tarski, there are an infinite number of different, mutually nonisomorphic models of ZF, including countable models and models of arbitrarily large cardinality (from the perspective of background set theory, see Section 5.4.6).

According to the results of Gödel and Cohen, the axiom of choice AC is undecidable in ZF. The same applies to the continuum hypothesis CH and the generalized continuum hypothesis GCH. Thus, among the ZF models, there are models of ZFC as well as models of ZF + ¬AC. Among the ZFC models, there are again an infinite number of different, mutually nonisomorphic models, for example, models of ZFC + GCH and models of ZFC + ¬GCH (cf. [64], p. 183). In each of these models, V exhausts the entire universe, i. e., the domain of the respective model.

The class **V** itself is not a model of ZFC because **V** (in contrast to the hierarchy levels V_α) is not a set. According to Gödel's second incompleteness theorem, no model of ZFC can be defined within ZFC, not even the existence of such a model can be proven (see [65], pp. 107–108). Within suitable extensions of ZFC, however, this is quite possible. If, for example, axioms that imply the existence of inaccessible cardinals are added to ZFC, and if κ is the smallest inaccessible cardinal, then V_κ is a model of ZFC ([106], p. 167).

Models for an axiom system of set theory are always taken from a presupposed set theory. This circumstance makes it necessary to distinguish between object set theory and background set theory and thus between different language levels.

5.4.5 Metalanguage and object language

In mathematical logic, we examine the language of mathematics with mathematical methods; and we use language for this purpose, namely (also) the language of mathematics. The threatening circularity of this procedure is broken through the distinction between language levels. Language thus appears in mathematical logic in (at least) two ways: on the one hand, as an object of investigation—then it is called *object language*—and on the other hand, as a tool or means of investigation—then it is called *metalanguage*.

Object languages are *formal* languages. What belongs to an object language is determined by its *alphabet* (a set of symbols) and its (recursively defined) *calculi of terms and formulas*. "Speaking" in an object language means specifying an object-language formula. Giving an object-language proof means giving a sequence of formulas according to the so-called *sequent calculus* (a system of inference rules).

Terms and formulas in an object language initially have no meaning. They only acquire a meaning through an *interpretation*, i. e., a function that assigns elements of the domain of a suitable structure to the constants and variables and corresponding functions or relations on the domain of this structure to the function and relation symbols of the language. An interpretation of a formula can lead to a true or a false statement. In the first case, the interpretation is called a *model* of the formula. If the formula does not contain any free variables (i. e., if it is a *sentence*), the structure used is also called a model of the formula.

The actual mathematics practiced by mathematicians is usually not formulated in a formal object language, but in a colloquial language (enriched with technical terms), however, in foundational discussions it is usually assumed that mathematics could be formulated in the framework of set theory and thus *in principale* in a first-order object language.

5.4.6 Background set theory and object set theory

In introductions to mathematical logic, set theory is usually used naively (i. e., without an axiomatic basis). Models are taken from a universe of sets whose existence is assumed to be given. ZFC provides an axiomatic theory for these considerations.

In fact, the entire first-order logic (syntax and semantics) can be built on the basis of ZFC (for example, with the elements of ω as a set-theoretic replacement for the variables and with suitable n-tuples for encoding terms, formulas, sequents, assignments, interpretations, and structures). The model relationship can be defined on the basis of ZFC using $\in$-formulas. The completeness theorem and other model-theoretical theorems can be symbolized by $\in$-formulas and derived from ZFC (see [65], Section VII.4).[25]

---

25 The authors point out (with reference to [19]) that a much weaker axiom system than ZFC is sufficient

Just as language is sometimes used as a tool (metalanguage) and sometimes as an object of investigation (object language), set theory also presents itself to us in two ways: Once as a tool (for syntactic and semantic considerations) and once as an object of investigation. In the first case, it is called *background set theory*, in the second case *object set theory*.

If the two language levels are conflated, this quickly leads to paradoxical statements. As an example, we will briefly discuss the *Skolem's paradox* (cf. [65], p. 107 f.): Provided that ZFC is consistent, ZFC has a model $\mathfrak{A}$ with countable domain A according to the Löwenheim–Skolem theorem. On the other hand, an ϵ-sentence can be derived from ZFC, which states that there is an uncountable set U and thus uncountable many objects in the universe of discourse A. How does this fit together?

The contradiction is resolved by separating the concepts of object set theory (labeled with superscript $\mathfrak{A}$ below for clarity) and background set theory. The uncountability$^{\mathfrak{A}}$ of U means that there is no element of A that is the interpretation of an injection$^{\mathfrak{A}}$ of U into $\omega^{\mathfrak{A}}$. This does not contradict the statement that in background set theory there is an injection of $\{x \in A \mid x \in^{\mathfrak{A}} U\}$ into ω. Thus, U is uncountable$^{\mathfrak{A}}$ from an *internal* point of view (in the sense of object set theory), but countable from an *external* point of view (in the sense of background set theory).

For a set theory realist, the universe of background set theory is the *true* set universe. In it, every ϵ-sentence (even one that is undecidable in ZFC) has a definite truth value, is either true or false. Axiom systems such as ZF and ZFC (and possibly extensions of them) capture essential properties of the true set universe, even if they cannot be complete. Within background set theory, different models of ZFC (as object set theory) can be investigated.

For a formalist, the set universe is a fiction and set theory is a formal theory without a semantic counterpart, without referents. Model-theoretical considerations take place within this fiction, trusting that the axioms of the background set theory (usually ZFC) are consistent.

For a constructivist, a set universe based on ZFC is not a meaningful object of investigation.

5.4.7 The so-called standard models

The natural numbers or the real numbers cannot be uniquely characterized (up to isomorphism) by their arithmetic axioms alone (all of which can be formulated in a first-order language). In a first-order language, the induction axiom for the natural numbers or the completeness axiom for the real numbers must be replaced by an axiom schema,

---

to prove the completeness theorem (cf. [65], p. 109). As mentioned in Section 5.1.7, the completeness theorem can even be derived in the finitistically reducible theory WKL_0.

which, however, does not guarantee uniqueness. According to the theorem of Löwenheim, Skolem, and Tarski, these axiom systems have models of arbitrarily large cardinality.

In a Zermelo–Fraenkel background set theory, the set of natural numbers (including zero) is identified with ω (see Section 5.2.2). The set ω is the smallest inductive set, i. e., the smallest set that contains $0^\omega := \emptyset$ and with each x also its successor $x \cup \{x\}$ (where the operations $\cup$ and $\{.\}$ are defined by $\in$-formulas). It exists according to the infinity axiom.[26]

With the recursion theorem, the addition $+^\omega$ and the multiplication $\cdot^\omega$ can be defined recursively as usual. Then the structure $(\omega, 0^\omega, +^\omega, \cdot^\omega)$ is a model of first-order *Peano arithmetic* (short: PA), i. e., a model of the axioms

$$\forall x \, \neg x + 1 = 0, \tag{5.7}$$

$$\forall x \, x + 0 = x, \tag{5.8}$$

$$\forall x \, x \cdot 0 = 0, \tag{5.9}$$

$$\forall x \forall y \, (x + 1 = y + 1 \to x = y), \tag{5.10}$$

$$\forall x \forall y \, x + (y + 1) = (x + y) + 1, \tag{5.11}$$

$$\forall x \forall y \, x \cdot (y + 1) = x \cdot y + x, \tag{5.12}$$

and the induction schema

$$\varphi(0) \wedge \forall x \, (\varphi(x) \to \varphi(x + 1)) \to \forall x \, \varphi(x) \tag{5.13}$$

for each $\{0, +, \cdot\}$-formula $\varphi(x)$ (possibly with parameters) (cf. [65], p. 169).

One calls $\mathfrak{N} := (\omega, 0^\omega, +^\omega, \cdot^\omega)$ the *standard model of arithmetic*. Nonstandard models of arithmetic were first investigated by Skolem ([203]).

Starting from ω, the number system up to the real numbers can be constructed in one of the known ways (see, for example, [66]). The set $\mathbb{R}$ constructed in this way, together with the constants $0^\mathbb{R}$ and $1^\mathbb{R}$ (defined by $\in$-formulas), the binary functions $+^\mathbb{R}$ and $\cdot^\mathbb{R}$, as well as the binary relation $<^\mathbb{R}$, is a model of the field axioms, the order axioms and the following axiom schema (which formalizes the least-upper-bound property for sets defined by φ):

$$\exists x \, \varphi(x) \wedge \exists b \forall x \, (\varphi(x) \to x \leq b)$$
$$\to \exists g \, (\forall x \, (\varphi(x) \to x \leq g) \wedge \forall c \, (\forall x \, (\varphi(x) \to x \leq c) \to g \leq c))$$

for each $\{0, 1, +, \cdot, <\}$-formula $\varphi(x)$ (possibly with parameters) (cf. [25], p. 356). Then $\mathfrak{R} := (\mathbb{R}, 0^\mathbb{R}, 1+^\mathbb{R}, \cdot^\mathbb{R}, <^\mathbb{R})$ is the *standard model of real arithmetic*. Nonstandard models are, for example, corresponding structures with domain ${}^*\mathbb{R}$.[27]

---

[26] In a class set theory without infinity axiom, ω can be defined as a proper class (see, for example, [27]).
[27] Structures of hyperreal numbers, however, do much more, since they are elementary equivalent to

Standard models are always taken from an assumed background set theory. But what about set theory itself? If ZFC is consistent, there is a model of ZFC according to Gödel's completeness theorem. However (as already mentioned in Section 5.4.2), the existence of such a model is not provable within ZFC, but only in some proper extensions of ZFC. In particular, there is no standard model of ZFC within ZFC. Switching to a more comprehensive set theory (for example, with axioms of large cardinals) would only postpone the problem, as there would then be no standard model for the more comprehensive set theory.

If one wants to grant the standard models an objective and absolute existence, one must start with an extra-mathematical postulate that demands the existence of an objective and absolute universe of sets in which the standard models (and all other models) are at home. This is the position of the set theory realists. It is also the working hypothesis of Shapiro's "working realist" (see Section 5.1.4). The universe of background set theory is perceived as *real* and can be equated with the intuitively plausible class **V** due to the axiom of regularity (see Section 5.4.2). It is the familiar set universe with all the familiar sets such as $\mathbb{N}_0$ (or ω), $\mathbb{Z}$, $\mathbb{Q}$, $\mathbb{R}$, and so on. "Familiar" is not a mathematical term. It refers to what the "working realist" perceives as real.

For a formalist, the universe of sets is a fiction. This means that all standard models are also fiction. They have no referents. Robinson puts it like this:

> I will mention here the assumption that there exists a standard or intended model of Arithmetic or (alternatively, but relatedly) of Set Theory. Clearly, to the formalist, the entire notion of standardness must be meaningless, in accordance with our first basic principle ([180], p. 242).

"Our first basic principle" refers to point (i) from the Robinson quote on page 141, i.e., to the statement that infinite wholes exist neither really nor ideally.

5.4.8 Nonstandard pictures of set theory

In the Sections 3.5 and 3.6, we presented several conservative extensions of ZFC, that made it possible to distinguish between standard and nonstandard sets by means of an additional predicate. The question is: What happens with these extensions to the "familiar set universe", i.e., to the reality of the "working realist"? Interestingly, one can take different perspectives on this (cf. [75], pp. 212–214).

Internal picture: The familiar universe of sets is understood as the universe of the extended theory, whose description merely becomes richer through the extension. Nothing is added to the universe, one can only talk about the universe in a more differentiated way. This perspective is taken by Nelson in his internal set theory (see Section 3.5).

---

the structure of real numbers with respect to a language that contains a symbol for *every* real number and for *every* relation on $\mathbb{R}$ (see Section 3.3.3).

Standard picture: The familiar universe of sets is understood as a subuniverse (the class of standard sets) of a larger universe of internal sets. All infinite standard sets receive additional (fictitious) nonstandard elements in the larger universe. This perspective is taken in [105].

External picture: The familiar universe of sets is understood as a subuniverse (the class of well-founded sets) of a larger universe that contains both internal and external sets. This picture is favored by Kanovei (see [110]).

The internal picture and the standard picture do not differ in terms of their mathematics, which in both cases takes place in the universe of internal sets. In fact, from a formalistic point of view (from which all set universes are fictions), there is no difference at all between the two pictures. However, the chosen picture may play a role in the acceptance of the nonstandard theory. In [105], it is reported that the textbook was only accepted when the internal picture was changed to the standard picture.

For many realists (or working realists), it is apparently easier to accept that fictitious elements are added to familiar sets than to assume that familiar sets contain elements that were previously unsuspected.

However, another note is important for realists. In Section 3.6, we explained that BST and HST are not only conservative extensions of ZFC, but that they can even be interpreted in ZFC. Put simply, this means that any ZFC universe can (by an appropriate interpretation) be reinterpreted as a BST or HST universe, in which the old ZFC universe can be found as the class of standard sets (in BST) or as the class of well-founded sets (in HST). In this sense, BST or HST are just as real as ZFC. The decision as to whether there should be nonstandard objects in the real universe of sets is thus also optional for ZFC realists. One could object that such an interpretation is possible in ZFC, but that it does not reflect the "true conditions" in the set universe. However, such an argument is not common in mathematics. Here is a comparison: In set theory, we also take the liberty of assuming the existence of urelements, if necessary, because this can be modeled in ZFC (see, for example, [106], p. 250).

5.4.9 Multiverse theories

According to the formalist position, we can only know about the set universe what can be deduced from the stipulated axioms. The concept of set is only fixed to the extent that the axioms allow. There is no absolute concept of set. This formalistic relativism has its realistic counterpart in the multiverse theories (for an overview, see, for example, [4]). What they have in common is that they are not based on a single set universe, but on a multitude of set-theoretical universes, all of which exist and each instantiate different set concepts. The diversity of models (see Section 5.4.4) is, to a certain extent, transferred to the level of universes.

Hamkins gives several principles (*Multiverse Axioms*) that postulate the existence of certain universes relative to others [88]. For example, universes exist if they are definable or interpretable in another universe (*realizability principle*) or if they emerge from another universe by forcing (*forcing extension principle*). The principles formalize that there is no particular universe that we can regard as *the* absolute background universe. For example, according to the *countability principle*, every universe is countable from the point of view of another universe or, according to the *well-foundedness mirage*, not well-founded from the point of view of another universe. Within ZFC, Gitman and Hamkins constructed a so-called *toy model* of their multiverse [79]. As there is no absolute concept of a set according to this multiverse theory, Hamkins also doubts that we have an absolute concept of finiteness, as this is linked to the natural numbers defined within set theory [87]. We will discuss the natural numbers in more detail in Section 5.5.

One can discuss whether a multiverse realism is more satisfactory than the formalist position, according to which there are different formal set theories (and thus set concepts), but no corresponding set universes. In any case, an absolute set concept is questionable from both a formalist and a realist position.

5.4.10 The role of the axiom of choice

Apart from the axiom of extensionality and the axiom of regularity, each of which expresses fundamental characteristics of the concept of set, the axioms of ZF are *existence axioms*. They require the existence of certain sets (possibly depending on parameters). Furthermore, they are *effective* in the sense that a certain instance of the required sets can be *defined* in ZF: For the infinity axiom, the set ω is definable. For the axioms of power set, union or pairing the respective operations $\mathcal{P}$, $\bigcup$ or $\{.,.\}$ are definable. The sets that exist according to the axiom schemas of separation or replacement can also be defined.

The situation changes fundamentally if we add the axiom of choice AC. It differs from the other axioms of existence in that the existence it requires remains indeterminate. In a common formulation, AC reads: For every family of nonempty and pairwise disjoint sets, there is a set that contains exactly one element of each set of the family. An alternative formulation is: For each family of nonempty sets, there is a *choice function*, i. e., a function that maps each set of the family to an element of that set. In general, however, it is not possible to explicitly specify a certain choice function for a given family of nonempty sets, i. e., to define it by an $\in$-formula. In particular, there is no definable operation that assigns a specific choice function to each family of nonempty sets.[28]

---

[28] If one adds the constructibility axiom $V = L$ to ZFC, the definition of such an operation is quite possible because there is a definable well-order predicate on L (see Section 5.4.3), and one can define the operation, for example, in such a way that the first choice function according to the well-order is taken.

Mathematical derivations that use the axiom of choice to prove the existence of a set without explicitly defining the set are usually called *inconstructive* or *nonconstructive*. However, it should be noted that these terms are used differently in constructivism. There, constructing a set means specifying an effective procedure with which one can (potentially) construct the elements of the set and distinguish them from one another (cf. Section 5.1.5 and the explanations on constructivism in Section 6.1.2). Constructivists therefore also reject the classical *axiom of power set* as inconstructive.

For analysis, countable versions of the axiom of choice are usually sufficient, such as the axiom of *countable choice* CC or the axiom of *dependent choice* DC. CC reads: For every *countable* family of nonempty sets, there is a choice function. DC reads: If R is a binary relation on a nonempty set A and for every $a \in A$ there is a $b \in A$ with bRa, then there is a sequence (a_n) with $a_{n+1}Ra_n$ for all $n \in \omega$. ZF proves the following implications (see [106], p. 50):

$$AC \Rightarrow DC \Rightarrow CC.$$

The inconstructive character of the axiom of choice carries over to the theorems that are proved with this axiom. Typically, they provide mere statements of existence without any possibility of determining one of the objects whose existence they ensure. In the basic lectures, for example, the axiom of choice is used to prove

- that every vector space has a basis,
- that continuity in terms of sequences implies ε–δ-continuity,[29]
- that every countable union of countable sets is countable again.[30]

CC is sufficient for the last two examples. In set theory, for example, AC is used to show that every set is equipotent to a cardinal number and thus two sets are always comparable with respect to their cardinality. However, the axiom of choice also has counterintuitive consequences such as the existence of non-Lebesgue-measurable subsets of $\mathbb{R}$ or the Banach–Tarski paradox (see [222]).[31] Its role in applied mathematics is therefore controversial.

According to the results of Gödel and Cohen, AC is independent of ZF (see Section 5.4.3). This means that if ZF is consistent, then both ZFC and ZF plus $\neg$ AC are consistent. For the history of the axiom of choice, see, for example, [157].

The axiom of choice is equivalent over ZF to the well-ordering theorem (every set can be well-ordered) and to Zorn's lemma (any partially ordered set, for which each chain has an upper bound, contains a maximal element) (see, for example, [64],

---

[29] The reverse implication can already be proven in Z^0 (ZF without regularity and without replacement) (see [64], pp. 113–114).

[30] The proof uses the conclusion that an enumeration can be *chosen* for each of the countably many sets.

[31] A narrative presentation of the Banach–Tarski paradox can be found in [123].

pp. 117–121). A comprehensive description of the role of the axiom of choice in mathematics can be found in [107] and [98].

Zorn's lemma is used to prove the ultrafilter theorem (every filter can be extended to an ultrafilter), which is employed to construct the hyperreal numbers and, more generally, nonstandard embeddings (see Sections 3.3.1 and 3.4.11). The reservations about the axiom of choice regarding its role in applied mathematics are therefore often transferred to nonstandard analysis (see Section 6.3). More precisely, over ZF the ultrafilter theorem is equivalent to the Boolean prime ideal theorem (BPI), which is weaker than AC, but does not follow from ZF alone ([107], p. 17). On the other hand, ZF+BPI is not strong enough to prove CC.

Is the axiom of choice indispensable for nonstandard analysis? The axiomatic approaches such as IST initially seem to offer a way out, as they do not require an ultrafilter construction and the axiom of choice could therefore be dispensed with. However, from the weakened theory ZF + I + S + T follows BPI and thus also the ultrafilter theorem (see [99]). Therefore, this theory is not conservative over ZF. On the other hand, for an elementary nonstandard analysis one does not need the full idealization and standardization from IST. A nonstandard theory that is conservative over ZF is presented in [104]. This is also the basis for the course outlined in Chapter 2.

5.4.11 Challenge: Nonstandard set universes

The familiar view
- For mathematics, a universe of background set theory is available that satisfies the axioms of ZFC (or at least ZF).
- This set universe can be identified with the class **V** of the cumulative hierarchy.
- It provides models for all consistent theories.
- Undecidable statements about sets have a definite truth value (realist position) or could be decided by additional axioms (formalist position). They have no relevance for applied mathematics.
- Finite sets are identified with naive finite collections.

Challenges posed by nonstandard analysis
- The universe of background set theory is not absolute.
- The conservative nonstandard extensions of ZFC give it unfamiliar properties while retaining all familiar properties.
- Familiar sets (such as $\mathbb{N}$, $\mathbb{R}$) contain unfamiliar elements (for example, unlimited numbers).
- The concept of finiteness is not absolute. In particular, the set-theoretical concept of finiteness must be distinguished from the naive concept of finiteness.

5.5 The natural numbers

Natural numbers are the oldest numbers we know. We use them as ordinal numbers to put things in order, as cardinal numbers when counting things, or simply for arithmetic.

The process of counting, which produces natural numbers, can be documented in the simplest way by the continued repetition of a particular sign (for example, tally marks on a piece of paper or notches on a bone). In many areas of everyday life, tally sheets are still a tried and tested means of representing small natural numbers, be it the number of drinks consumed, votes counted in an election, or scores in games and sports. For larger numbers or for arithmetic, we generally use the decimal system, which has become so commonplace that we often identify numbers with their decimal representation.

Mathematicians like to leave the question of what natural numbers actually are to the philosophers, while they themselves retreat to the axiomatic point of view. The spectrum of philosophical answers ranges from "natural numbers are real" to "natural numbers are nothing but the symbols used to represent them". But even the mathematical, i. e., axiomatic answer is not as clear-cut as is often assumed.

5.5.1 Peano structures

What we expect from the natural numbers in mathematics has been defined since Peano and Dedekind by the Peano–Dedekind axioms—usually just called *Peano axioms* for short. They uniquely characterize the natural numbers up to isomorphism, but cannot be formulated in a first-order language because of the induction axiom. If we use a second-order language with a constant 0, a unary function symbol σ (for the successor function) and a predicate variable X, the Peano axioms in modern notation look like this, for example (see [65], p. 47):

$$\forall x \forall y\, (\sigma x = \sigma y \to x = y), \tag{5.14}$$

$$\forall x\, \neg \sigma x = 0, \tag{5.15}$$

$$\forall X (X0 \land \forall x\, (Xx \to X\sigma x) \to \forall y\, Xy). \tag{5.16}$$

Axiom (5.14) expresses that the successor function is injective and (5.15) that zero is not in the range of the successor function. Axiom (5.16) is the second-order induction axiom: Any predicate X that applies to zero and that, if it applies to a number, always also applies to the successor of this number, applies to all numbers. The models of this second-order axiom system are precisely the *Peano structures*, i. e., the structures with a domain A and a function σ^A defined on A (the interpretation of σ) and an element 0^A of A (the interpretation of 0) such that:

(P1) $\sigma^A : A \xrightarrow{\text{inj}} A,$

(P2) $0^A \notin \mathrm{ran}(\sigma^A)$,
(P3) $\forall B \, [B \subseteq A \wedge 0^A \in B \wedge \forall x \, (x \in B \Rightarrow \sigma^A(x) \in B) \Rightarrow B = A]$.

Axioms (P1) to (P3) are the set-theoretic Peano axioms. According to Dedekind's theorem (see, for example, [65], p. 47), all Peano structures are isomorphic. For mathematicians, it is therefore irrelevant which Peano structure is used to define the natural numbers.

In ZFC, the natural numbers are usually defined as the elements of the smallest limit ordinal ω (see Section 5.2.2). With $0^\omega := \emptyset$ and $\sigma^\omega(z) := z \cup \{z\}$, $(\omega, 0^\omega, \sigma^\omega)$ is a Peano structure (see [64], pp. 67–68).

5.5.2 Object numbers and metanumbers

In an object set theory, which allows the definition of Peano structures and the derivation of Dedekind's theorem, the natural numbers of the object language (short: *object numbers*) must be distinguished from the natural numbers of the metalanguage (short: *metanumbers*). We explain this using the example of ZFC. For clarification, we use $\mathfrak{n}$ (instead of n) as a placeholder for metanumbers (as in Section 2.5.4).

With the commonly defined function symbols $\cup$ and $\{.\}$ define another function symbol σ by $\sigma x := x \cup \{x\}$. Then for each metanumber $\mathfrak{n}$ there is an object-language term

$$\sigma^\mathfrak{n} \emptyset := \underbrace{\sigma \cdots \sigma}_{\mathfrak{n} \text{ times}} \emptyset,$$

which denotes the $\mathfrak{n}$th successor of $\emptyset$. For every metanumber $\mathfrak{n}$, there is therefore a corresponding object number $\sigma^\mathfrak{n} \emptyset$. So far, the considerations are purely syntactic in nature and get by with potential infinity.

If one assumes a background set theory that allows for infinite sets and the derivation of Gödel's completeness theorem, model-theoretical considerations are possible. There are then models in background set theory for each consistent axiom system. Assuming the consistency of ZFC, let $\mathfrak{A}$ be a model of ZFC and A its domain. The interpretations of the symbol $\in$ and the symbols defined in the object language (such as $\emptyset, \omega, \sigma$) are denoted by the respective symbols with superscript $\mathfrak{A}$. Then each term $\sigma^\mathfrak{n} \emptyset$ has a unique interpretation in A, namely $(\sigma^\mathfrak{n}\emptyset)^\mathfrak{A} = (\sigma^\mathfrak{A})^\mathfrak{n}(\emptyset^\mathfrak{A})$ (the function $\sigma^\mathfrak{A}$ applied $\mathfrak{n}$ times to $\emptyset^\mathfrak{A}$). This means that each metanumber has a unique equivalent in the domain of the model.

Now, let $\mathcal{M}$ denote the set of metanumbers, i. e., the smallest inductive set in *background set theory*. The set

$$\mathcal{N} := \{(\sigma^\mathfrak{n}\emptyset)^\mathfrak{A} \mid \mathfrak{n} \in \mathcal{M}\}$$

is a subset of A. Since it is a construct of background set theory, it is not available in object set theory. Therefore, it cannot be used to define the set of natural numbers within object set theory. However, the set $\mathcal{N}$ can be compared with the set

$$\mathcal{N}' := \{a \in A \mid a \in^{\mathfrak{A}} \omega^{\mathfrak{A}}\}$$

in background set theory. The set $\mathcal{N}$ contains exactly the counterparts of the metanumbers in the model while $\mathcal{N}'$ contains exactly the object numbers, but viewed *externally* (as elements of the domain of the model). With the above designations, we have the following theorem of background set theory:

Theorem 59. *Let $\mathfrak{A}$ be a model of ZFC. Then $\mathcal{N}$ is a subset of $\mathcal{N}'$.*

Proof. Since $\mathfrak{A}$ is a model of ZFC, we have $\emptyset^{\mathfrak{A}} \in^{\mathfrak{A}} \omega^{\mathfrak{A}}$ and for all $a \in A$, if $a \in^{\mathfrak{A}} \omega^{\mathfrak{A}}$, then $\sigma^{\mathfrak{A}}(a) \in^{\mathfrak{A}} \omega^{\mathfrak{A}}$. By induction (in the background set theory), we have $(\sigma^n \emptyset)^{\mathfrak{A}} \in^{\mathfrak{A}} \omega^{\mathfrak{A}}$ for all $n \in \mathcal{M}$. Hence, $\mathcal{N}$ is a subset of $\mathcal{N}'$. □

However, $\mathcal{N}$ and $\mathcal{N}'$ are generally not the same. While it is true for all $a \in A$ that inductive$^{\mathfrak{A}}(a)$ implies $\omega^{\mathfrak{A}} \subseteq^{\mathfrak{A}} a$, it is not certain that $\mathcal{N}$ is an element of A at all. Therefore, the conclusion is not applicable to $\mathcal{N}$.

Put simply, there may be "more" object numbers than metanumbers. This possibility is realized, for example, in ZFC models that satisfy the extensions from Section 3.5 or 3.6.

5.5.3 Which are the true natural numbers?

The need to distinguish between the natural numbers of different language levels may tempt us to ask which are the "true" natural numbers. However, with such a question we leave the area of responsibility of mathematics, at least mathematics as it is understood today.

Formalized mathematics—or mathematics that claims to be formalizable in principle—always presupposes the natural numbers (if only to have an unlimited supply of variables). As long as one remains on the syntactic level and, for example, examines the term, formula, and sequent calculi with proof-theoretical means, one gets by with potential infinity (for example, on the basis of PRA). As soon as semantics comes into play, we need models, and that means we need a suitable background set theory (which in turn can be formalized)—or at least a sufficiently strong subsystem of Z_2, for example, WKL$_0$ (see Section 5.1.7). In these formal systems, the natural numbers are again object numbers and must be distinguished from the metanumbers (which, strictly speaking, are already metametanumbers).

What does this mean for the question in the section heading? The answer depends on the philosophical point of view.

The formalist position is finitistic with regard to metamathematics. This means that the claim to speak about an infinite totality of all metanumbers is not made at all. Accordingly, the potentially infinite metanumbers would be best described as the "true" natural numbers. Everything that goes beyond this leads to object numbers of a formal theory, which are to be distinguished from the metanumbers. The phenomenon of incompleteness must be accepted for both metanumbers and object numbers. This means that undecidable statements remain in any case.

Such an answer is not acceptable to realists. According to the realist view, every arithmetical statement must have an unambiguous answer, regardless of whether we can determine it within a formal theory. Wilholt adopts a pragmatic attitude towards questions of consistency or completeness as long as only formal systems are involved, but describes such an attitude as "unacceptable" ("nicht hinnehmbar", [227], p. 234) in relation to the core areas of classical mathematics (to which he particularly counts elementary arithmetic and real analysis).

We turn once again to his partial mathematical realism, according to which natural numbers are universals ante rem, i. e., number properties of factual or counterfactual aggregates (see Section 5.1.4). Starting from the (presumably not infinitely possible) physical aggregates, I can only recognize a *potential* infinity of counterfactual aggregates (which would arise if there were further causal processes that could be added). I do not see how this justifies talking about the totality of *all* number properties or about *any totality* of number properties, as is done in (P3) ([227], p. 182) ((P3) there is a version of Peano's set-theoretic induction axiom, i. e., (P3) in Section 5.5.1, but formulated for aggregates). The natural numbers derived from this partial realism therefore fit better with the finitistic view.

A set theory realist can equate the "true" natural numbers with the elements of the *real* set ω. However, even a realist cannot rule out that the real ZFC universe also satisfies the axioms of IST (or another conservative nonstandard extension of ZFC). According to the considerations from Section 5.4.8, there are at least the following options for the existence of natural nonstandard numbers:

- The nonstandard numbers can be assumed to be elements of ω according to the internal picture.
- The nonstandard numbers can be added to the real numbers as fictitious elements according to the standard picture.
- The nonstandard numbers can be regarded as real by interpreting BST (or HST) in ZFC.

We note that the existence of natural nonstandard numbers larger than any sum of ones cannot be ruled out from either a formalist or a realist position. The equation of the naive, intuitively given natural numbers with the elements of ω is not tenable.

5.5.4 Challenge: Unlimited natural numbers

The familiar view
- In (background) set theory, there is a Peano structure $(\omega, 0^\omega, \sigma^\omega)$. It is unique up to isomorphism.

- The elements of ω are by definition *finite* (i. e., the finite cardinals).
- The naive natural numbers are intuitively identified with the elements of ω.

Challenges posed by nonstandard analysis
- The naive natural numbers must be distinguished from the natural numbers of set theory (i. e. the elements of ω).
- The elements of ω are (by definition) the finite cardinals, but some of them can be larger than any sum of ones and in this sense "infinitely large" (unlimited).

6 From the foundations of mathematics: On the status of nonstandard numbers

In this chapter, we revisit the basic philosophical questions from Section 5.1.3 in relation to standard and nonstandard numbers. We restrict ourselves to the consideration of the hyperreal numbers in ZFC and the real numbers in the nonstandard extensions of set theory discussed in Sections 3.5 and 3.6. Firstly, these are particularly relevant for nonstandard analysis, since a powerful transfer principle applies to them, and secondly, they are particularly interesting with regard to the basic philosophical questions, because of the nonconstructive elements in the respective theories. Other number systems that contain infinitesimals, such as Laugwitz's omega numbers defined with the filter Cof (see Section 3.2) or Tall's superreal numbers (see Section 4.2.3) can be defined within ZF and are therefore considered less problematic. However, these number systems are not nearly as efficient due to the limited or missing transfer principle. Moreover, the omega numbers only form a partially ordered ring.

6.1 Ontological questions

So let us start with the ontological questions. Do nonstandard numbers exist in the same way as standard numbers or are they on a lower ontological level? Do they exist at all or are they even contradictory concepts? It is clear that ontological questions have to be discussed in relation to basic philosophical convictions. Contradiction, however, is not compatible with existence for any position in the philosophy of mathematics. We therefore ask first:
– Are nonstandard numbers contradictory?

And then:
– Do nonstandard numbers exist?

Here we examine realist, formalist, and constructivist positions, as described in Section 5.1.

6.1.1 Are nonstandard numbers contradictory?

Infinitesimal quantities
Quantities are the historical precursors of the real numbers and infinite and infinitesimal quantities are the historical precursors of the infinitely large and infinitely small nonstandard numbers.

As useful as infinitesimal quantities were in the new infinitesimal calculus of Newton and Leibniz, they were also controversial. From the very beginning, they had to contend with the accusation that they were an impossibility. A quantity that is nonzero at the beginning of a calculation and then miraculously becomes zero in the course of the calculation was bound to be a provocation. George Berkeley gleefully put his finger in the wound here when he criticized the vanishing increments "o" in Newton's fluxion calculus:

> I admit that Signs may be made to denote either any thing or nothing: And consequently that in the original Notation $x + o$, o might have signified either an Increment or nothing. But then which of these soever you make it signify, you must argue consistently with such its Signification, and not proceed upon a double Meaning: Which to do were a manifest Sophism ([32], p. 24).

Berkeley's mockery of the vanishing increments as "ghosts of departed quantities" ([32], p. 59) has become famous.[1]

Cantor also vehemently polemicized against infinitesimal quantities, for example, in his letter to Giulio Vivanti dated December 13, 1893, where he calls them the "infinite cholera bacillus" of mathematics and "paper quantities" that "have no existence other than on the paper of their discoverers and followers" ([154], pp. 505–506). Cantor was convinced that he had shown the impossibility of infinitesimal quantities. However, his proofs of impossibility were based on irreconcilable, but not compelling, assumptions about the systems of magnitudes. A detailed discussion of Cantor's arguments can be found in [69].

The modern answer

In today's nonstandard analysis, the seemingly contradictory attributes of infinitesimals are satisfactorily reconciled. Infinitesimals are not first nonzero and then zero; they are not variable or vanishing quantities, but fixed numbers, just like standard numbers. They are simply numbers whose *absolute value is smaller than any positive standard number*. Hence zero is the only infinitesimal standard number. Infinitesimals are (as with Leibniz) divisible. There is no smallest positive infinitesimal number.

The "generalized equality" is the equivalence relation "$\approx$". The miraculous "becoming zero" of an infinitesimal difference at the end of the calculation of a differential quotient is simply the transition to the standard part because the derivative is defined as the standard part of the differential quotient (analogous for the integral and other concepts of analysis).

A great achievement of nonstandard analysis is to explain precisely which statements about standard numbers can be transferred to the extended number system (by the transfer principle). In the Robinson approach, these are the first-order statements. In

---

[1] However, the founders of infinitesimal calculus did not actually claim that one and the same quantity was initially nonzero and later equal to zero, but rather that infinitesimal parts can be discarded in the final result, for example, in the sense of Leibniz's "generalized equality" (cf. Section 2.1).

the nonstandard set theories, the transfer principle corresponds to the transfer axiom, which applies to internal statements.

Today, consistency is usually understood as consistency relative to ZFC. In this sense, we can state that Infinitesimals (and nonstandard numbers, in general) are not contradictory.

6.1.2 Do nonstandard numbers exist?

The realist position

From ZFC follows the existence of elementary non-Archimedean field extensions of $\mathbb{R}$ and, more generally, the existence of nonstandard embeddings, as described in Section 3.4. For a ZFC realist, the existence of nonstandard numbers is therefore beyond question. For multiverse realists (see Section 5.4.9), the universes of the various nonstandard set theories, in which nonstandard numbers exist in $\mathbb{R}$, are also real.

Nonstandard numbers in $\mathbb{R}$ exist in various conservative extensions of ZFC (for example, IST). They can therefore also be regarded as real by ZFC realists (according to the internal picture) or as fictions added to the real universe (according to the standard picture). In addition, interpretable nonstandard extensions of ZFC (e. g., BST or HST) can be understood as "real" (in the sense of the interpretation) (see Section 5.4.8).

Similar statements apply to ZF realists and proponents of a partial realism who consider the standard models of $\mathbb{N}$ or $\mathbb{R}$ definable in ZF to be real: Since there are conservative nonstandard extensions of ZF (see note at the end of Section 5.4.10), nonstandard numbers can be considered real (according to the internal picture) or can be added to the real universe as fictions (according to the standard picture).

The constructivist position

For constructivists, existence means constructibility (in finitely many steps over the natural numbers). Infinite sets, functions, sequences do not exist as *objects*, but only in the sense of an explicitly specifiable procedure that allows a *potentially* infinite construction. Bedürftig and Murawski refer to this as "restricted existence" ([26], p. 167).

According to Erret Bishop (cf. [39], p. 67), one defines a *set A* by specifying what to do to construct an element of A and what to do to show that two elements of A are *equal* (in the sense of an equivalence relation $=_A$). For example, to construct an element of the set $\mathbb{Q}$, form an ordered pair (p, q) with $p, q \in \mathbb{Z}, q \neq 0$, and define

$$(p, q) =_\mathbb{Q} (p', q') \quad :\Leftrightarrow \quad pq' =_\mathbb{Z} p'q.$$

You can proceed in the same way beforehand to define $\mathbb{Z}$ on the basis of $\mathbb{N}$. If the context is clear, the lower index at the equality sign is omitted for simplification.

To define an *operation f* from A to B (for two sets A, B), one must specify what is to be done to obtain a construction of $f(a) \in B$ for a given construction of $a \in A$. If $a = a'$

always results in $f(a) = f(a')$, f is called a *function* or *mapping*. A *sequence* is a function with domain $\mathbb{N}$.

Before we come to the question of the existence of nonstandard numbers, it is helpful to realize in what sense the real numbers or the set $\mathbb{R}$ exist in constructive analysis. We refer here again to [39], p. 18.

A sequence (x_n) of rational numbers is called *regular* if

$$|x_m - x_n| \leq m^{-1} + n^{-1} \quad (m, n \in \mathbb{N}). \tag{6.1}$$

A *real number* is a regular sequence of rational numbers. Two real numbers $x \equiv (x_n)$ and $y \equiv (y_n)$ are *equal* if

$$|x_n - y_n| \leq 2n^{-1} \quad (n \in \mathbb{N}). \tag{6.2}$$

The "equality" between real numbers defined in this way is an *equivalence relation*.

Let us look at the simple statement $\sqrt{2} \in [1, 2]$. According to the realist or formalist view, this is a statement about two related (real or fictitious) objects: the real number $\sqrt{2}$ and the closed real interval $[1, 2]$. Both are entities in a (real or fictitious) universe (for example, a ZFC universe). Not so in constructivism. There, as we have seen above, infinite sets do not exist as objects. The symbols $\sqrt{2}$ and $[1, 2]$ are meaningless in themselves. A meaning only arises when they are used within a statement such as $\sqrt{2} \in [1, 2]$. However, this is merely an abbreviated notation for a statement about natural numbers.

In constructivism, statements about the existence of something (in definitions or theorems) must always include the construction of what is said to exist. The definition of continuity then looks like this, for example:

Definition 30 ([39], p. 38). A real-valued function f defined on a compact interval I is *continuous* on I if for each $\varepsilon > 0$ there exists $\omega(\varepsilon)$ such that $|f(x) - f(y)| \leq \varepsilon$ whenever $x, y \in I$ and $|x - y| \leq \omega(\varepsilon)$. The operation $\varepsilon \mapsto \omega(\varepsilon)$ is called a *modulus of continuity* for f.

In contrast to the classical ε–δ-definition, the construction of a suitable δ for a given ε must therefore be explicitly ensured by an operation (the modulus). This corresponds to the general principle of constructivist mathematics: existence means constructibility. Based on this definition, it can be shown constructivistically, for example, that the function $x \mapsto x^2$ is continuous on $[0, 1]$. Other results of classical analysis, such as Bolzano's theorem for continuous functions with a change of sign, do not hold in constructivism, as a zero cannot generally be constructed. Instead, it can only be shown that a real number exists (i. e., can be constructed) whose function value is arbitrarily small (see [39], p. 40).

For Bishop, the meaning of a statement lies in its *computational content*, and he sees such a meaning neglected in formalist mathematics. In this respect, he even accuses nonstandard analysis of a "debasement of meaning". In *Crisis in Contemporary Mathematics*, he writes:

> My interest in non-standard analysis is that attempts are being made to introduce it into calculus courses. It is difficult to believe that debasement of meaning could be carried so far ([38], pp. 513–514).

Bishop's review of Keisler's textbook on nonstandard analysis, *Elementary Calculus*, is therefore correspondingly critical:

> Although it seems to be futile, I always tell my calculus students that mathematics is not esoteric: It is common sense. (Even the notorious ε, δ definition of limit is common sense, and moreover is central to the important practical problems of approximation and estimation.) They do not believe me. In fact the idea makes them uncomfortable because it contradicts their previous experience. Now we have a calculus text that can be used to confirm their experience of mathematics as an esoteric and meaningless exercise in technique ([36], p. 208).

Sam Sanders deals with the accusations of Bishop and Connes that nonstandard analysis lacks computational content due to its nonconstructiveness.[2] In his article *The unreasonable effectiveness of Nonstandard Analysis*, Sanders provides a *template*, called $\mathcal{CI}$, which can be used to convert a pure nonstandard theorem and its proof into a constructive version. The term "pure" here means that the nonstandard definitions are used for continuity, derivative, integral, etc. More precisely, theorems and proofs are converted from the systems P or H, which are conservative extensions of Peano or Heyting arithmetic. In each case, the extension is a fragment of Nelson's internal set theory.

Sanders describes how his template works as follows:

> Intuitively speaking, the 'effective version' of a mathematical theorem is obtained by replacing all its existential quantifiers by functionals providing the objects claimed to exist. In other words, the object claimed to exist by the theorem at hand can be computed (in a specific technical sense) from the other data present in the theorem ([189], p. 462).

The expression "computed (in a specific technical sense)" here means a calculation via primitive recursive functions in the sense of Gödel's system T (see [84]).

According to Sanders, the template $\mathcal{CI}$ is applicable to most theorems of "ordinary" (i. e., not set-theoretic) mathematics, as understood in reverse mathematics on the basis of the "Big Five" (see Section 5.1.7). In detail, the investigations are technically demanding. However, the results show that the statements of nonstandard analysis do indeed have a computational content (and thus a "meaning" in Bishop's sense).

In constructivism, neither the real numbers nor the nonstandard numbers exist as objects in the way that natural numbers do.

Further explanations of constructivist nonstandard analysis can be found, for example, in [165] and [166] (nonstandard models are constructed in Martin-Löf's construc-

---

[2] Connes understands "constructive" classically from the point of view of ZF (see Section 6.2.1), whereas Bishop understands it from the point of view of the constructivist foundational program. Both aspects are treated by Sanders.

tive type theory) and [31] (a version of Nelson's internal set theory that is compatible with Bishop's constructive analysis). Walter Schnitzspan had already developed a constructive nonstandard analysis based on Weyl and Lorenzen's constructive analysis in his dissertation [191].

The formalist position

The formalist position is the least problematic in terms of ontology, since mathematical objects only claim a formal existence within the framework of a theory that is assumed to be consistent. The usual assumption in mathematics today is that ZFC is consistent. Since the axiom of choice is independent of ZF, it is unproblematic with regard to the question of consistency. Thus, from a formalistic point of view, hyperreal numbers have the same right to exist as everything whose existence can be derived from ZF. As conservative extensions of ZFC, the nonstandard set theories IST, BST, HST deserve the same confidence as ZFC.

To be or not to be, is that the question?

How relevant are ontological questions for mathematics? Is it important that mathematical objects exist in a certain sense? This is, of course, not a mathematical question. Modern formalistic mathematics does not provide ontological answers. Behrends points out in the preliminary remarks of his analysis textbook that mathematics does not investigate what *is*, but what can be *deduced* (cf. [28], p. 5). From such a deductivist position, there are no ontological objections to nonstandard numbers, because their existence follows in conservative extensions of widely accepted theories such as ZF or ZFC.

The statement with which Leibniz defended the use of infinite and infinitesimal quantities in his Calculus as early as 1676 (and which we had already quoted in excerpts in Section 2.1) sounds astonishingly modern:

> The things we have said up to now about infinite and infinitely small quantities will appear obscure to some, as does anything new; nevertheless, with a little reflection they will be easily comprehended by everyone, and whoever comprehends them will recognize their fruitfulness. Nor does it matter whether there are such quantities in the nature of things, for it suffices that they be introduced by a fiction, since they allow economies of speech and thought in discovery as well as in demonstration. Nor is it necessary always to use inscribed or circumscribed figures, and to infer by reductio ad absurdum, and to show that the error is smaller than any assignable; although what we have said in Props. 6, 7 & 8 establishes that it can easily be done by those means. Moreover, if indeed it is possible to produce direct demonstrations of these things, I do not hesitate to assert that they cannot be given except by admitting these fictitious quantities, infinitely small or infinitely large (see above, Scholium to Prop. 7) ([138], p. 128, as translated by Arthur in [9], p. 27).[3]

---

3 In the original Latin: "Quae de infinitis atque infinite parvis huc usque diximus, obscura quibusdam videbuntur, ut omnia nova; sed mediocri meditatione ab unoquoque facile percipientur: qui vero per-

In this respect, Leibniz often compared infinitesimals with imaginary roots (for example, in [136]). In [137] he called them *well-founded fictions* ("des fictions bien fondées").

Proponents of ZF realism or empirically grounded realism (see Section 5.1.4) could deny the real existence of nonstandard numbers, but not the possibility of introducing them as well-founded fictions (in the context of conservative extensions). That is all that modern mathematics demands. And Leibniz did not demand more either.

6.1.3 Summary of the ontological answers

Nonstandard numbers are elements of conservative extensions of ZF and ZFC. In this sense, they are not contradictory. They exist for ZFC or multiverse realists as hyperreal numbers or as nonstandard numbers in $\mathbb{R}$. Their existence cannot be ruled out in ZF or empirically grounded realism either. At the very least, nonstandard numbers can be assumed to be useful and well-founded fictions.

For constructivists, neither real nor hyperreal numbers exist as objects.

For formalists, the question of metaphysical existence does not arise. Nonstandard numbers exist (like the real numbers) within the framework of suitable formal theories that are assumed to be consistent (and whose consistency relative to ZF is certain).

6.2 Epistemological questions

What can we know about nonstandard numbers or nonstandard objects? Assuming their existence, can we know anything at all about them or say anything concrete about them? And if so, is our knowledge of nonstandard objects just as certain (or just as uncertain) as our knowledge of standard objects? We discuss the following questions:
- How definite is nonstandard analysis?
- How reliable is nonstandard analysis, foundationally speaking?

6.2.1 How definite is nonstandard analysis?

Important real numbers (such as $\sqrt{2}$, π, e), as well as the set $\mathbb{R}$ of all real numbers, are concrete and defined objects for most mathematicians (for those who accept ZF or

---

ceperit, fructum agnoscet. Nec refert an tales quantitates sint in rerum natura, sufficit enim fictione introduci, cum loquendi cogitandique, ac proinde inveniendi pariter ac demonstrandi compendia praebeant, ne semper inscriptis vel circumscriptis uti, et ad absurdum ducere, et errorem assignabili quovis minorem ostendere necesse sit. Quod tamen ad modum eorum quae prop. 6. 7. 8. diximus facile fieri posse constat. Imo si quidem possibile est directas de his rebus exhiberi demonstrationes, ausim asserere, non posse eas dari, nisi his quantitatibus fictitiis, infinite parvis, aut infinitis, admissis, adde supra prop. 7. schol."

ZFC as a basis). They can be *defined* in the language of set theory, i. e., (in principle) by ϵ-formulas based on the ZF axioms.[4] More precisely:
- There are ϵ-formulas that define the symbols $\mathbb{R}$, 0^R, 1^R, $+^R$, $\cdot^R$, $<^R$ on the basis of ZF in such a way that ZF proves: $(\mathbb{R}, 0^R, 1^R, +^R, \cdot^R, <^R)$ is a complete ordered field.
- ZF also proves: All complete ordered fields are uniquely isomorphic to $(\mathbb{R}, 0^R, 1^R, +^R, \cdot^R, <^R)$.
- There are ϵ-formulas that define concrete irrational numbers.

Corresponding statements do not hold for $^*\mathbb{R}$ or nonstandard numbers in $^*\mathbb{R}$. The construction of the hyperreal numbers (and, more generally, of the associated nonstandard world) uses a δ-incomplete ultrafilter $\mathcal{U}$ on a suitable set J (see Section 3.4.11). Although the existence of such an ultrafilter (for a fixed J) follows from ZFC (where the axiom of choice is essential), it is by no means unique. Due to the nonconstructiveness of the existence proof, it is also not possible to arbitrarily determine a specific ultrafilter. Therefore, $^*\mathbb{R}$ is not determined in the same way as $\mathbb{R}$ is.

The extension $^*\mathbb{R} \supset \mathbb{R}$ is not canonical

The extensions of the number systems $\mathbb{C} \supset \mathbb{R} \supset \mathbb{Q} \supset \mathbb{Z} \supset \mathbb{N}$ are each motivated by certain closure requirements, either algebraic or, in the case of $\mathbb{R} \supset \mathbb{Q}$, order-theoretic. The extensions are *canonical* in the sense that they are (up to isomorphism) the unique or at least the most parsimonious answer to the motivating problem. The set $\mathbb{Z}$ is the smallest ring that includes $\mathbb{N}$, $\mathbb{Q}$ is the smallest field that includes $\mathbb{Z}$, $\mathbb{R}$ is *the* Dedekind complete extension field of $\mathbb{Q}$, and $\mathbb{C}$ is *the* algebraic closure of $\mathbb{R}$.

In contrast, the transition from $\mathbb{R}$ to a non-Archimedean, elementary equivalent extension field $^*\mathbb{R}$ is *not* canonical. There are infinitely many such extensions with essentially different properties, for example, arbitrarily large cardinality. Therefore, different constructions of $^*\mathbb{R}$ are generally not isomorphic. Bell and Machover state:

> We therefore regard nonstandard analysis as an important tool of clarification, exposition and research – often beautiful, sometimes very powerful, but never exclusive ([30], p. 573).

Assuming the continuum hypothesis CH (i. e., $\mathfrak{c} := 2^{\aleph_0} = \aleph_1$), it can be shown that all hyperreal fields of cardinality $\mathfrak{c}$ (and thus all hyperreal fields constructed with a nonprincipal ultrafilter on $\mathbb{N}$) are isomorphic (see Theorem 3.5 in [73]). Without the assumption of CH, the situation is undetermined.

---

[4] For the concept of *definition* of constants, function, and relation symbols in formal languages, see, for example, [65], Section VIII.3.

Definable models

The general indeterminacy of the field extension $^*\mathbb{R} \supset \mathbb{R}$ does not rule out that there are *definable* models and that models with certain additional properties are uniquely determined (up to isomorphism). We quote a result by Kanovei and Shelah, provable in ZFC,

Theorem 60 ([111], Theorem 1). *There exists a definable, countably saturated[5] extension $^*\mathbb{R}$ of the reals $\mathbb{R}$, elementary in the sense of the language containing a symbol for every finitary relation on $\mathbb{R}$.*

and a more general

Theorem 61 ([111], Section 4, items 1 and 2). *For any given infinite cardinal κ, there is a κ-saturated elementary extension of $\mathbb{R}$, definable with κ as the only parameter of definition.[6] There are special elementary extensions of $\mathbb{R}$, of as large cardinality as desired. Special models of the same cardinality are isomorphic.*

The term κ-*saturated* here means that every family $\mathcal{D}$ of internal subsets of $^*\mathbb{R}$ with the finite intersection property and cardinality strictly less than κ has a nonempty intersection (cf. Definition 19 in Section 3.4.10). *Countably saturated* means the same as $\aleph_1$-*saturated* (where $\aleph_1$ is the least uncountable cardinal), i. e., that every *countable* family $\mathcal{D}$ of internal subsets of $^*\mathbb{R}$ with the finite intersection property has a nonempty intersection.

Theorem 61 makes a statement about so-called *special models*. A model $\mathfrak{A}$ with domain A is called *special* if it is the union of an elementary chain of models $\mathfrak{A}_\beta$ with cardinals $\beta < \text{card}(A)$, where the $\mathfrak{A}_\beta$ are each β^+-saturated (and β^+ denotes the next larger cardinal after β). According to a theorem in model theory, special models of the same cardinality are isomorphic (cf. [47], pp. 292–295).

Are there definable nonstandard numbers?

Unless otherwise stated, we understand *definable* here to mean *definable by an $\in$-formula in the background set theory ZFC*, where certain models of the hyperreals (such as the Kanovei–Shelah model $^*\mathbb{R}$ in Theorem 60) can be defined.

---

[5] For a cardinal κ, κ-saturation means that every family $\mathcal{D}$ of internal subsets of $\mathbb{R}$ with the finite intersection property and cardinality strictly less than κ has a nonempty intersection. Countable saturation means $\aleph_1$-saturation (where $\aleph_1$ is the least uncountable cardinal), see Section 3.4.10.

[6] Definable here means *ordinal-definable* (see [106], p. 194). A set X is called *ordinal-definable* if there is a $\in$-formula φ and ordinal numbers $\alpha_1, \ldots, \alpha_n$ with $X = \{u \mid \varphi(u, \alpha_1, \ldots, \alpha_n)\}$. Ordinal-definability can be expressed in the language of set theory, for it can be shown that the ordinal-definable sets are precisely the elements of the class $OD = \bigcup_{\alpha \in \text{Ord}} \text{cl}\{V_\beta \mid \beta < \alpha\}$, where cl denotes the *Gödel closure*, i. e., the closure under the *Gödel operations* $G_1, \ldots, G_{10}$ (see [106], p. 177). Similarly, it can be expressed in the language of set theory that a set X is *ordinal-definable from A* (where A is a set), i. e., definable by an $\in$-formula with ordinals and A as parameters (see [106], p. 195).

In nonstandard analysis, not a single nonstandard hyperreal is concretely defined (by an ∈-formula in ZFC). With regard to a construction of the hyperreals using a nonprincipal ultrafilter, it is possible to specify representing sequences for certain nonstandard numbers, for example, the sequence $(1, 2, 3, \ldots)$ for "the" hypernatural number Ω, but which number this is (for example, whether it is even or odd) depends on the ultrafilter, and a concrete nonprincipal ultrafilter cannot be specified. Strictly speaking, you would have to include the ultrafilter in such designations of hyperreal numbers and write $[(1, 2, 3, \ldots)]_\mathcal{U}$, for example. This number is just as indeterminate as the ultrafilter $\mathcal{U}$.

So are there any definable nonstandard numbers at all? Alain Connes argues as follows: Any given nonstandard number produces in a canonical way a non-Lebesgue-measurable subset of the real interval $[0, 1]$. However, it follows from work by Cohen and Solovay that it is impossible to explicitly produce such a subset (cf. [53], p. 14).

> So, what this says is that for instance in this example, nobody will actually be able to name a non standard number. A nonstandard number is some sort of chimera which is impossible to grasp and certainly not a concrete object (ibid).

Connes is apparently referring here to the so-called *Solovay model*, a model of ZFC, in which it is true that every set of reals definable from a countable sequence of ordinals (and thus also every definable set of reals) is Lebesgue measurable. Solovay used Cohen's forcing method to show that such a model exists, provided that there is a model of ZFC + "There is an inaccessible cardinal" (see Theorem 2 in [204]). A definable nonstandard number that produces a definable non-Lebesgue-measurable subset of $[0, 1]$ would contradict the properties of the Solovay model. Further details of Connes' argument and a rebuttal to his general criticism of Robinson's nonstandard analysis can be found in [109].

A related argument for why there cannot be definable nonstandard numbers in $^*\mathbb{R}$ refers to nonprincipal ultrafilters.[7] ZFC proves the existence of nonprincipal ultrafilters on ω, but in ZFC we *cannot* prove that there is a nonprincipal ultrafilter on ω that is *definable* (see Theorem 1 in [171]).[8] Let $^*\mathbb{R}$ be a definable model (which exists according to Theorem 60) and suppose there is a definable nonstandard number $t \in {}^*\mathbb{R}$. We can assume that t is infinite (otherwise replace t with $(t - \text{stp}(t))^{-1}$). Now define $k := [|t|]$, where $|.|$ and $[.]$ are (the $*$-extensions of) the absolute value function and the greatest integer function. Then k is an infinite hypernatural and

$$\mathcal{U}_k := \{S \in \mathcal{P}(\omega) \mid k \in {}^*S\}$$

---

[7] I owe the reference to this argument to Ali Enayat.

[8] Theorem 1 in [171] even contains the stronger statement that nonprincipal ultrafilters on ω are not ODR, i. e., are not even definable with ordinal and real parameters. The theorem makes no use of inaccessible cardinals.

a nonprincipal ultrafilter on ω and definable. Contradiction! Therefore, $^*\mathbb{R}$ cannot contain a definable nonstandard number.

On the other hand, there are (relatively) consistent extensions of ZF in which concrete nonstandard numbers and non-Lebesgue-measurable subsets of $\mathbb{R}$ are definable. For example, from ZF plus Gödel's constructibility axiom $\mathbf{V} = \mathbf{L}$ follow the axiom of choice and the existence of an explicitly specifiable well-ordering of all sets (see Section 5.4.3). This means that a concrete nonprincipal ultrafilter can be defined (e.g., the first according to the well-ordering) and thus that a concrete extension $^*\mathbb{R} \supset \mathbb{R}$ and concrete nonstandard numbers can also be defined.

In Nelson's internal set theory (see Section 3.5), there is no need for an ultrafilter construction of $^*\mathbb{R}$ because the nonstandard numbers can be found in the definable set $\mathbb{R}$, but a concrete nonstandard number cannot be specified there either. Although the idealization axiom ensures the existence of nonstandard numbers, it does not allow us to pick out a specific one. All objects that can be defined by internal predicates are standard due to the transfer axiom. Set definitions with external predicates are only permitted in conjunction with the standardization axiom, which also only produces standard sets. In fact, it can be proven that any set definable by an ϵ-st-formula in IST (possibly with standard parameters) is standard (see Theorem 1 in [108]).[9]

How important is determinacy?

As explained above, the field extension $^*\mathbb{R} \supset \mathbb{R}$ is not canonical. However, this need not be seen as a disadvantage. On the contrary, you can choose a suitable field extension for different purposes.[10]

It is not necessary to have a concrete, definable model for $^*\mathbb{R}$ in order to use nonstandard numbers in analysis. It is sufficient to know that suitable models exist. For elementary analysis, any ultrafilter construction is suitable; for advanced applications, a sufficiently saturated nonstandard embedding is required.

It is also not necessary to define specific nonstandard numbers because nonstandard numbers are always used in a qualitative sense. It is never important to use a definable nonstandard number. In the proof of the intermediate value theorem, for example, any infinitesimal number could be used to decompose the interval. And the nonstandard definitions of continuity, derivative, and integral are of the type "if for all $h \approx 0, \dots$". Even in a set theory in which a specific nonstandard number can be defined (for example, in ZF + ($\mathbf{V} = \mathbf{L}$)), the use of explicitly defined nonstandard numbers could be dispensed with without loss, as they have no added value for analysis.

Machover admits that the nonstandard definition, for example, of continuity, is more intuitive and easier to use than the classical ε–δ-definition and that it is therefore

---

[9] I thank Vladimir Kanovei for this reference.
[10] In [109], a parallel is drawn in this respect to the extension of $\mathbb{Q}$ to algebraic number fields. This is also not canonical. Different extensions can be useful for different purposes.

tempting to use only the nonstandard definition in calculus courses, but that this could not replace the standard definition, as the invariance of the chosen extension would have to be shown and this could only be done in a simple way by showing the equivalence to the standard definition ([146], p. 208).

However, Kanovei, Katz, and Mormann rightly point out that the use of nonstandard analysis in calculus courses is not about replacing standard with nonstandard definitions, but rather about facilitating the introduction to the conceptual world of calculus through the more intuitive nonstandard definitions. It is therefore not a question of either/or, but of the chronological order ([109], p. 32).

Furthermore, an introduction to calculus focuses on working on an axiomatic basis. Questions concerning the construction of models or the categoricity of the axiom system are initially of secondary importance in methodological and educational terms.

6.2.2 How reliable is nonstandard analysis, foundationally speaking?

The short answer is: relatively reliable (i.e., as reliable as standard analysis). In Section 3.5, it was mentioned that IST is conservative over ZFC: Every internal sentence that is provable in IST is also provable in ZFC. In particular, this means that IST is consistent relative to ZFC: If ZFC is consistent (which is generally assumed), then so is IST. Since the axiom of choice is independent of ZF, IST is also consistent relative to ZF.

Furthermore, according to [104]: The ZF-extension SPOT is conservative over ZF, and this is provable in ZF. The conservativeness can be formulated (with suitable coding) as a Π_2^0-sentence in WKL_0 and can therefore even be proved finitistically, i.e., in PRA (cf. Definition 25 in Section 5.1.7).

6.2.3 Summary of the epistemological answers

- The field extension $^*\mathbb{R} \supset \mathbb{R}$ is not unique or canonical, but there are definable models for the hyperreals.
- In nonstandard analysis, no specific nonstandard number is defined, but it is sufficient to know that nonstandard numbers with certain properties (e.g., $x \approx 0$ or $n \gg 1$) exist. Further determination is not important for the usefulness of nonstandard numbers.
- Nonstandard analysis is just as reliable as standard analysis.

6.3 Questions on applicability

With *real numbers*, the name says it all. They are used to numerically capture what is apparently real – the continuous phenomena in space and time. In contrast, the hyper-

real numbers seem to overshoot the mark, even by their very name; they seem to be more than is needed to describe the real world. We therefore ask:
– What can nonstandard numbers mean for the real world?

And then:
– Is nonstandard analysis applicable to the real world?

In addressing these questions, we are particularly interested in whether there is a difference between standard numbers and nonstandard numbers in terms of their meaning for the real world, or whether there is a difference between standard analysis and nonstandard analysis in terms of their applicability to the real world. By the "real world" we mean here, in summary and in general, the object of investigation of the empirical sciences.

6.3.1 What can nonstandard numbers mean for the real world?

Nonstandard numbers obviously cannot be the result of a physical measurement. The same is true for irrational standard numbers. Physical measuring devices always cover a limited measuring range and only allow a limited measuring accuracy. With digital measuring devices, both depend on the number of digits displayed; with analog measuring devices, they depend on the size of the display field and the fineness of the reading scale. A physical measurement is ultimately nothing more than a count of units.

So is there any difference at all between irrational numbers and nonstandard numbers in terms of their meaning for the real world? Imagine an apparatus with which we can continuously change a physical quantity, for example, an applied voltage. Let the voltage gradually increase from 0 volts (at time t_0) to 1 volt (at time t_1). Now compare the following statements:
– There is a point in time between t_0 and t_1 at which the applied voltage was $\frac{\pi}{4}$ volts.
– There is a point in time between t_0 and t_1 at which the applied voltage was infinitesimal.

Which of these statements is physically meaningful? If one assumes that physical quantities (possibly also space and time) are quantized, one can question both statements because irrational or infinitesimal voltages then have no physical meaning. In a classical physical description, both statements make sense in relation to a *mathematical model* that contains irrational and infinitesimal numbers.

The analysis in the previous sections (especially Sections 5.3 to 5.5 and 6.1) has shown that equating the physical or intuitive continuum with the mathematical model of real numbers is not permissible. The continuum can be modeled Archimedically (by $\mathbb{R}$) or non-Archimedically (by the various models of $^*\mathbb{R}$). And even if we opt for the model $\mathbb{R}$, infinitesimals are not ruled out. The real numbers are a construct of set theory and whether there are unlimited and infinitesimal ones among them depends on the assumptions we make about the set universe from which we take the real numbers. It therefore does not change anything if we introduce the real numbers axiomatically in a second-order language because the models of this theory are defined within a set theory in which metanumbers and object numbers are to be distinguished (see Section 5.5.2). In no case can the existence of nonstandard numbers in $\mathbb{R}$ be excluded. You can only decide not to talk about nonstandard numbers. For this reason alone, equating (positive) real numbers with ratios of physical magnitudes (ante rem) is problematic.

The significance of the real numbers in IST[11] or the hyperreal numbers in ZFC for the real world lies in the fact that they are elements of a mathematical structure with which many phenomena of the real world can be modeled well, i.e., theoretical predictions are sufficiently well confirmed by observations. This applies to nonstandard analysis at least to the same extent as it applies to standard analysis because all the results of standard analysis can be reproduced in nonstandard analysis. Furthermore, nonstandard analysis offers additional modeling options (see Section 6.3.3).

In practice, i.e., as actual measurement results, neither irrational standard numbers nor nonstandard numbers occur. In theory (for example, in formulas that are intended to model real-world phenomena), however, both irrational standard numbers (for example, π, e, $\sqrt{2}$) and nonstandard numbers can occur. The special feature of nonstandard numbers in this context is that they are only determined qualitatively, for example, as $\xi \approx 0$ or $\nu \gg 1$ (see the examples in Section 6.3.3). The fact that no nonstandard number can be denoted by a defined constant has already been discussed in Section 6.2.1. The nonstandard numbers share this fate with almost all standard numbers.

6.3.2 Is nonstandard analysis applicable to the real world?

The quintessence of the last section was that $\mathbb{R}$ or $^*\mathbb{R}$ can only be applied to the real world as set-theoretically constructed models and that the derivability of the existence of nonstandard numbers in $\mathbb{R}$ depends on the presupposed set theory. In ZFC, the axiom of choice is crucial for the construction of $^*\mathbb{R}$. IST also contains the axiom of choice and even IST without the axiom of choice implies the ultrafilter theorem (see Section 5.4.10). The axiom of choice is known to have counterintuitive consequences, for example, the existence of non-Lebesgue-measurable subsets of $\mathbb{R}$ and the Banach–Tarski paradox. One can therefore ask whether models whose proof of existence depends on the axiom of choice are at all suitable to be applied to the real world. Even books on nonstandard analysis contain skeptical statements in this regard, for example, Väth:

---

[11] The following considerations apply analogously to the nonstandard extensions of ZFC discussed in Section 3.6.

> However, the use of nonstandard analysis has the drawback that even the simplest results make use of the axiom of choice ...
> The above observation is an essential disadvantage since this means in the author's opinion that nonstandard analysis is not a good model for "real world" phenomena ([224], p. 85).

In contrast, Landers and Rogge state that "in the applied sciences in particular, it has been shown that the nonstandard domain $^*\mathbb{R}$ is often better suited for modeling than the classical domain $\mathbb{R}$ of real numbers" ([127], p. 2).[12] So what is true? The question is when a model is a *good* model for real-world phenomena. Must the elements of the model have referents in the real world? If this were the case, then, as we have seen, the model of real numbers would also have to be questioned. Does a good model have to be independent of the axiom of choice? In contrast to the construction of hyperreal numbers, the construction of real numbers does not require the axiom of choice. Nevertheless, many applications of standard analysis cannot do without the axiom of choice (at least in its countable version).[13] Moreover, there are conservative extensions of ZF that provide nonstandard numbers and a transfer axiom, as discussed in Section 5.4.10. It is therefore not true that nonstandard analysis necessarily depends on the axiom of choice.

Nonstandard analysis is indisputably used successfully in the empirical sciences, be it in economics [141], financial mathematics [7], or physics [61, 7]. Mathematical modelling (standard or nonstandard) in the empirical sciences is almost always accompanied by idealizations. The strength of nonstandard analysis often lies in the fact that (in addition to the standard modeling) another type of idealization is available, which is in some way closer to reality and easier to handle, for example by replacing a very large number (for example, of market participants, atoms, spatial regions, or events) with an infinitely large number N (a nonstandard number), which can nevertheless be treated as a finite number.[14] In Section 6.3.3, we look at an example from probability theory.

We emphasize once again that the real numbers (as well as the hyperreal numbers) are a construct of set theory and that it is problematic to equate them with physical phenomena (cf. Section 6.3.1). Even if it were true that certain standard numbers correspond to physical length ratios, as some variants of realism postulate (see Section 5.1.4), while nonstandard numbers are only idealizations within mathematical models, there is no reason not to apply these models to the real world or to consider them less applicable. The usefulness is justification enough. The dependence of hyperreal numbers on

---

[12] In the original German: "Gerade in den angewandten Wissenschaften hat sich gezeigt, daß der Nichtstandard-Bereich $^*\mathbb{R}$ zur Modellbildung häufig besser geeignet ist als der klassische Bereich $\mathbb{R}$ der reellen Zahlen."

[13] In [20], the authors point out that, for example, the σ-additivity of the Lebesgue measure is not provable in ZF (p. 851).

[14] The different nature of the idealization in nonstandard analysis sometimes leads to misunderstandings. See, for example, the objections in [228, 63, 173, 174, 167] and the rebuttals in [20, 40, 41].

the axiom of choice does not contradict this, nor does the statement that the extension $^*\mathbb{R} \supset \mathbb{R}$ is not canonical or that no concrete nonstandard number can be defined.

It was already clear to Leibniz that the question of applicability had to be separated from the question of metaphysical existence. In a letter of 1716 to Samuel Masson, he emphasizes that infinitesimal calculus is useful when it comes to applying mathematics to physics, but that he does not claim to explain the nature of things with it because he considers infinitesimal quantities to be useful fictions.[15]

6.3.3 An example from probability theory

Imagine a random experiment with a spinner that randomly and without preference chooses any point on a circle. For the sake of simplicity, we identify the circle with the interval $[0, 1[$. A typical modeling of such an experiment is a continuous uniform random variable X on the probability space $(\Omega, \mathcal{A}, P)$, where the sample space Ω is the real interval $[0, 1[$, $\mathcal{A}$ is the Borel σ-algebra on Ω and P is the Borel measure. In this model, each elementary event (a singleton $\{\omega\}$, with $\omega \in \Omega$), has the probability 0. Due to the σ-additivity of P, every countable subset of Ω also has the probability 0.

In a discrete modeling one can use, for example, the sample space

$$\Omega_n := \left\{ \frac{k}{n} \mid k = 0, \ldots, n-1 \right\}$$

(for a fixed $n \in \mathbb{N}$) and the probability space $(\Omega_n, \mathcal{P}(\Omega_n), P_n)$ with

$$P_n(A) = \frac{|A|}{|\Omega_n|} = \frac{|A|}{n}, \quad \text{for } A \subseteq \Omega_n.$$

Each elementary event $\{\omega\}$ with $\omega \in \Omega_n$ then has the probability $\frac{1}{n}$. Here, only the impossible event $\emptyset$ has the probability 0.

Any physical experiment with a material spinner can be adequately described with such a discrete model (and sufficiently large n), because the measurement accuracy of physical experiments is limited. Continuous modeling abstracts from physical limitations and thus is an idealization of the real situation.

However, continuous modeling is not the only possible idealization. In the hyperreal numbers, modeling by the probability space $(\Omega_\nu, {}^*\mathcal{P}(\Omega_\nu), P_\nu)$ is possible, where ν is an infinite hypernatural number.

The following relationship exists between the continuous "real" modeling and the discrete "hyperreal" modeling:

---

[15] In the original French: "Le calcul infinitesimal est utile, quand il s'agit d'appliquer la Mathematique à la Physique, cependant ce n'est point par là que je pretends rendre compte de la nature des choses. Car je considere les quantités infinitesimales comme des fictions utiles" ([135], p. 629).

$$P(A) \approx P_\nu(^*A \cap \Omega_\nu) \quad \text{for each } A \in \mathcal{A}.$$

Alexander Pruss has objected (see [174]) that hyperreal probabilities are *underdetermined* in the following sense:
1. Singletons cannot be assigned a unique infinitesimal probability.
2. For each hyperreal-valued probability measure, there are an uncountable number of others that lead to the same decision-theoretical preferences (and are thus equivalent for modeling physical random experiments).[16]

However, Botazzi and Katz have pointed out that the probability of a singleton is uniquely determined as soon as the hyperfinite sample space has been fixed (in the above example, the hypernatural number ν) and that the probability measure is uniquely determined if it is assumed to be internal [41].[17] This is a natural assumption in Robinson's nonstandard analysis. In fact, $P(A)$ is well defined as a hyperfinite sum $\sum_{\omega \in A} P(\omega)$ only for internal P.[18]

If one does not model with hyperreal numbers, but in IST, the restriction to internal measures and events does not appear at all, since all sets are internal. In his book *Radically Elementary Probability Theory* [162], Nelson develops a probability theory with minimal means from his internal set theory that is so powerful that even demanding topics such as Brownian motion can be tackled. This probability theory is radically elementary because it remains entirely finite and does not require measure and integration theory. A *finite probability space* is a pair (Ω, P) consisting of a finite sample space Ω and a function $P : \Omega \to \mathbb{R}^+$ with $\sum_{\omega \in \Omega} P(\omega) = 1$. An *event* is any subset of Ω. The probability of an event A is defined as $P(A) = \sum_{\omega \in A} P(\omega)$.[19] A *random variable* is any function $X : \Omega \to \mathbb{R}$. The *expected value* of a random variable X is defined as $E(X) = \sum_{\omega \in \Omega} X(\omega) P(\omega)$. A *stochastic process* is a mapping $\xi : T \to \mathbb{R}^\Omega$ with a finite index set T. For each $t \in T$, $\xi(t)$ is then a random variable (often denoted by X_t). For example, if $T = \{1, \ldots, \nu\}$ and X is a random variable on (Ω, P), then (Ω^ν, P^ν) with

$$P^\nu(\omega_1, \ldots, \omega_\nu) := \prod_{n=1}^{\nu} P(\omega_n)$$

---

[16] For a given probability measure P, Pruss defines a probability measure P_α for each positive $\alpha \in \mathbb{R}$ by $P_\alpha(A) = \operatorname{Std} P(A) + \alpha \operatorname{Inf} P(A)$, where Std and Inf denote the standard part and the infinitesimal part of finite hyperreal numbers, respectively. With his Theorem 1, he specifies the decision-theoretic equivalence of P and all P_α.

[17] Pruss raises an analogous objection for probability measures that fulfill the Ω-limit axiom. The argument in [41] against this objection is also analogous. The underdetermination disappears if one restricts oneself to internal probability measures.

[18] This does not mean that only internal measures are considered in Robinson's non-standard analysis (cf. [85], Chapter 16, on the topic of *Loeb measures*).

[19] The function defined by this on $\mathcal{P}(\Omega)$ is again denoted by P, since there is no risk of misunderstanding.

a finite probability space, and the random variables X_n with

$$X_n(\omega_1, \ldots, \omega_\nu) := X(\omega_n), \quad 1 \leq n \leq \nu,$$

are independent. This stochastic process describes the ν-fold independent repetition of a random experiment given by X. In Nelson's radical elementary probability theory, a nonstandard number can now be chosen for ν in order to formulate, for example, the laws of large numbers ([162], Chapter 16).

In conventional probability theory, an infinite sequence of random variables $(X_n)_{n \in \mathbb{N}}$ is assumed to formulate the laws of large numbers. Consider the effort required to define infinite products of probability spaces and to show the existence of infinite families of independent random variables (cf. [21], § 9). Even if Ω contains only two elements (for example, when modeling a coin toss), $\Omega^\mathbb{N}$ is an uncountable sample space. Each elementary event of the product space (the result of an infinitely repeated coin toss) has probability 0. Only certain (measurable) subsets of $\Omega^\mathbb{N}$ are events. The probability of an event is no longer the sum of the probabilities of the elementary events covered, but an integral, and so on.

It is crucial to note that both conventional modeling with $(X_n)_{n \in \mathbb{N}}$ and nonstandard modeling with $X_1, \ldots, X_\nu$ ($\nu \gg 1$) are an idealization compared to physical experiments. Nelson writes:

> The conventional approach involves an idealization, because one cannot actually complete an infinite number of observations. The second approach also involves an idealization, because one cannot actually complete a nonstandard number of observations. In fact, it is the nature of mathematics to deal with idealizations ([162], p. 13).

In both cases, the idealization consists of replacing a quantitative term ("large") that is dependent on the application context with a qualitative theoretical term ("infinitely large" or "unlimited"). The difference lies in the concept of infinity used in each case (see Section 5.2).

6.3.4 Summary of the answers on applicability

- Both $\mathbb{R}$ and $^*\mathbb{R}$ are set-theoretically constructed models and idealizations when applied to the real world.
- Idealizations can be based on different concepts of infinity. In addition to infinite sets, nonstandard analysis also provides infinitely large and infinitely small numbers for idealization.
- Whether the existence of nonstandard numbers in $\mathbb{R}$ (and $\mathbb{N}$) can be derived depends on the set theory used.
- There are conservative extensions of ZF that provide nonstandard numbers and a transfer principle.

- Nonstandard numbers (if they occur explicitly in models of real-world phenomena) are only determined qualitatively (e. g., as $x \approx 0$ or $n \gg 1$).
- The usefulness of nonstandard numbers in modeling real-world phenomena does not depend on their ontological status.

7 Conclusion

7.1 Summary of the results

7.1.1 Nonstandard analysis in higher education

The textbooks that are usually recommended as literature in the lectures on analysis consistently take a classical approach based on Weierstrass's limit concept (see Section 4.1). With a few exceptions, nonstandard analysis is not mentioned in these textbooks. Infinitesimals are mentioned, if at all, in historical notes.

The survey discussed in Section 4.4 confirmed the picture reflected in the textbooks: Nonstandard analysis plays virtually no role in the introduction to calculus at university (in Germany). None of the 50 lecturers who took part in the survey used elements and methods of nonstandard analysis in the Analysis I or II lectures.

7.1.2 The teachers' assessment

The teachers' responses on their assessment of the suitability of nonstandard analysis for teaching reveal a more differentiated picture when different types of courses are considered (see Section 4.4.2). Overall, it can be stated that although the lecturers who gave an assessment were almost unanimous in their view that nonstandard analysis is unsuitable for the basic calculus lectures, almost one in three were at least positively disposed towards its use in complementary courses (e. g., undergraduate seminars).

Arguments for the use of nonstandard analysis included aspects of cognitive advantage (promoting understanding, development of intuition, more elegant and intuitive proofs), the opportunity to repeat material and the promotion of awareness of mathematical foundations. The main counterarguments were unfavorable conditions (lack of time in basic lectures, deviation from content specifications, few suitable textbooks, lack of human resources, and lack of competence among teachers), excessive demands or confusion among students (due to a high degree of abstraction and a lack of prior knowledge), the low relevance for mathematics and the lack of benefit (low added value compared to standard analysis, not of use in the further course of the study) (see Section 4.4.3).

7.1.3 On the negative attitude towards nonstandard analysis

To analyze possible reasons for a negative attitude towards nonstandard analysis in teaching, we focused on the response categories that indicate possible conflicts with habits of thought or values, rather than on the response categories that indicate unfavorable conditions.[1]

Based on the arguments *lack of prior knowledge, excessive demands on students, low relevance for mathematics* and *lack of benefit* mentioned in the survey, the following possible reasons for a negative attitude towards a use of nonstandard analysis in teaching emerged in Section 4.4.4:
- conflict with the value of "mathematical rigor" in university teaching,
- conflict with recognized educational principles (do not confuse or overwhelm, no inappropriately high abstraction right at the beginning),
- conflict with the desire to teach something relevant,
- conflict with the desire to teach something useful.

The relatively frequent mention of the counterargument *low relevance* (in 14 of 29 interviews) prompted us to examine in Chapter 5 the typical image of mathematics conveyed in a mathematics course for habits of thought that are particularly challenged by nonstandard analysis and can therefore lead to a negative attitude among teachers who act as mediators of this image of mathematics. The habits of thought primarily concern the concepts of infinity, number and continuum as well as the role of set theory.

Infinite sets are self-evident objects of mathematics from the very beginning, while infinitely large numbers do not appear at all at first, later possibly as infinite ordinal and cardinal numbers or as $\pm\infty$ in the extended real number system $\mathbb{R}$. Such infinite numbers are not field elements and do not lead to infinitesimals. In analysis textbooks, it is pointed out that the symbols dx and dy in the differential quotient $\frac{dy}{dx}$ are only historical designations and must not be used independently for numbers (see, for example, [28], p. 247). This note is quite appropriate in standard analysis, but may leave the impression that infinitely small numbers are dubious per se. Although non-Archimedean field extensions of $\mathbb{Q}$ or $\mathbb{R}$ are easy to produce by adjunction, they hardly play a role in studies, just like more complicated non-Archimedean fields, and so it is quite possible to finish a mathematics degree without ever having come into contact with infinitely small numbers.

The axiomatic-deductive character of mathematics is also emphasized in mathematics studies from the very beginning. We recall the quote from Tall on page 130 about the "*consequence* of advanced mathematics, based on abstract entities which the individual must construct through deductions from formal definitions" ([214], p. 20, emphasis in original). But the axioms that stand at the beginning of deduction do not come from nowhere. The completeness axiom in analysis (for example, as the axiom of Dedekind completeness) is a transfer of the (historically relatively young) Cantor–Dedekind view of the continuum, i. e., the concept of the linear continuum as a set of extensionless

---

[1] The reported institutional resistance to experimental calculus courses was also caused by a negative attitude or a suspected negative attitude of individuals towards nonstandard analysis (cf. Section 4.3.4).

points (including all rationals), in which each division into a left and a right part is generated by a unique point. As a consequence, the continuum is isomorphic to $\mathbb{R}$, in particular it is Archimedean and uncountable. It is easy to get the impression that the axioms of the real numbers *describe* the intuitive linear continuum (i. e., the geometric straight line), but they rather *postulate* properties of an abstract set whose elements are then mentally projected onto the geometric straight line. How abstract the real numbers are only becomes clear when you actually construct them set-theoretically with all the intermediate steps. Hardly anyone imagines this construction of multiple nested infinite sets when thinking of real numbers. This construction is usually wisely omitted in analysis lectures.

Even if set theory and logic lectures are not compulsory courses as part of a mathematics degree, such a degree usually conveys a certain understanding of mathematics that is based on set theory and logic and that has become the *common sense* of modern mathematics. This includes the assumption of the existence of a universe of sets that can be described by suitable axiom systems (e. g., ZFC) and that serves as a reservoir for models of consistent theories. This universe of background set theory is thus to a certain extent the "reality" behind the official formalism, something mathematicians rely on when they are not themselves working on conceptions for the foundations of mathematics (see Section 5.4). In this background set theory, which is perceived as real, there is a unique complete ordered field $\mathbb{R}$ (up to isomorphism) and a unique Peano structure ω, of which an isomorphic copy is found as the smallest inductive subset of $\mathbb{R}$, which is then usually denoted $\mathbb{N}_0$ (or $\mathbb{N}$). Therefore, it is obvious (not in an ontological, but in a structuralist sense) to regard the elements of $\mathbb{N}$ as *the* natural numbers and (due to the continuum concept described above) $\mathbb{R}$ as *the* continuum.

Nonstandard mathematics can provoke resistance because it forces us to question the usual image of mathematics. In addition to the dominant Cantorian concept of infinity (and the improper numbers $\pm\infty$), there is another concept of infinity with completely different properties (see Section 5.2.2). Infinitely large and infinitely small numbers are suddenly self-evident objects of mathematics. Arithmetic and cardinal infinities stand side by side on an equal footing and claim to be equally useful concepts for mathematics.

The new arithmetic concept of infinity also affects the concept of the continuum. The linear continuum can no longer simply be identified with the real number line. Rather, it is advantageous not to imagine the continuum as a point set from the outset, but to understand it (as before the invention of set theory) as a *guiding idea* in its own right, independent of the concept of set.[2] Set-theoretical constructions such as $\mathbb{R}$ or $^*\mathbb{R}$ can reproduce certain aspects of this guiding idea and are in this sense *models* of the geometric continuum, but not identical with it.

---

[2] Hermann Weyl called the continuum a *medium of free emergence* (as translated by Scholz in [192]), in the German original, "Medium freien Werdens" ([226], p. 49).

Edward Nelson's internal set theory (IST) poses a particular challenge (see Section 3.5), since as a conservative extension of ZFC it preserves the mathematical "reality" described above, on the one hand, but, on the other hand, makes familiar objects appear unfamiliar. IST demonstrates in a particularly clear way the need to distinguish between different language levels and between theoretical and naive concepts. The set-theoretical *finite* is not the naive *finite*, the elements of $\mathbb{N}_0(:= \omega)$ are not the naive natural numbers. This is true in ZFC as well as in IST, but in IST the difference is explicit and obvious. The set $\mathbb{N}_0$ contains nonstandard numbers that are greater than any sum of ones, $\mathbb{R}$ contains positive numbers that are less than the inverse of any sum of ones. Every infinite set contains nonstandard objects. On the other hand, there is a finite set that contains *all* standard objects (including all sums of ones). Such apparent paradoxes are difficult to reconcile with the familiar picture of mathematics.

And as if that were not imposition enough, the *internal picture* suggested by IST (see Section 5.4.8) requires us to assume that all the nonstandard objects already existed in our familiar ZFC universe, but were not *seen* by us. In IST (and related theories), *making visible* is achieved through language extension (see Section 3.5.1).

7.1.4 Critical reflection on the reasons for rejection

The analysis in Section 4.4 has shown that the mathematically or didactically justified rejection of nonstandard analysis in teaching is essentially based on teachers' prejudices. This concerns both the feared dilemma of either overburdening students or having to compromise on mathematical rigor, as well as the perception of teaching something useless or irrelevant.

With a single additional postulate, the elementary extension principle, nonstandard proofs of calculus can be carried out with the same mathematical rigor as standard proofs (see Section 4.4.5). No inappropriately abstract concepts (formal logic, model theory, or ultrafilters) are required. Based on previous experience with experimental calculus courses, students are neither confused nor overwhelmed by a nonstandard introduction. On the contrary, there are indications that such an introduction accommodates students' existing conceptualizations and promotes intuitive understanding of the basic concepts (see Sections 4.3.1, 4.3.2 and 4.3.3). The feedback from the students themselves is also positive. For students of natural sciences, engineering, or economics, there is the additional advantage that the nonstandard approach is closer to the reasoning used in these subjects when applying calculus.

A general benefit of nonstandard mathematics lies in the broadening of horizons (for example, with regard to the foundations of mathematics and the historical perspective) and in the enrichment of the range of methods. The latter is also the main reason for the relevance of nonstandard analysis within and outside mathematics (see Section 4.4.6).

In addition to the didactic reasons for rejection, which could be read directly from the teachers' responses in the survey, there is great potential for resistance to nonstandard analysis (in teaching and in general) in the habits of thought that go hand in hand with the standard picture of mathematics described above and which are challenged by nonstandard analysis (see Section 7.1.3). These habits of thought include the primacy of Cantor's concept of infinity, which manifests itself in the axiom of infinity and which made modern mathematics based on set theory possible in the first place. However, Cantor's view is not the only possible and not the original way of thinking the actual infinite in mathematics. We recall that Leibniz allowed infinite quantities as fictions, but rejected infinite multiplicities because they violate the part–whole axiom (see Section 5.2.2). Since both concepts of infinity coexist without problems in ZFC or in conservative extensions of ZFC and are fruitful for mathematics, there is no reason for a one-sided preference for Cantor's infinity.[3]

The second relevant habit of thought concerns the primacy of standard models, associated with the Cantor–Dedekind postulate (the equation of the linear continuum with $\mathbb{R}$) and the standard model hypothesis, i. e., the implicit assumption that the elements of the smallest inductive set in background set theory correspond to the intuitively given natural numbers (see Sections 5.3.3, 5.4.7, 5.5.3). However, such a view is not tenable. The Cantor–Dedekind postulate is, as the name suggests and as described above, only a (nonmandatory) *postulate* and not a fact, and the equation of object numbers (of an axiomatic background set theory) and (intuitively given) metanumbers is not permissible (see Section 5.5.2). The prominent position of $\mathbb{N}$ and $\mathbb{R}$ (as domains of certain standard models) always arises only within the framework of a given background set theory. It cannot be derived solely from an empirically grounded realism (which, for example, understands natural numbers as universals of physical aggregates and positive real numbers as universals of physical proportions). Such numbers taken from experiential reality lead (even with counterfactual extrapolation) at most to potentially infinite number systems (see Sections 5.5.3 and 6.3.1). Infinite models, which are considered in the context of a transfinite background set theory, always require the distinction between metanumbers and object numbers and thus also the possibility of nonstandard numbers (see Section 5.5.2).

Reservations about nonstandard analysis are often justified by the fact that the extension $^*\mathbb{R} \supset \mathbb{R}$ is not canonical, requires the axiom of choice, and does not allow even a single nonstandard number to be explicitly specified (see Section 6.2.1). The hyperreal numbers thus seem to be inferior to the real numbers ontologically, epistemologically and in terms of their applicability to real world phenomena. However, the ex-

---

3 It is even possible to develop nonstandard analysis in "weak theories" that do not include Cantor's infinity at all (see, for example, [10]). Another interesting result by Harvey Friedman says that a particular theory *PA of nonstandard numbers, together with the so-called $^*\Pi_\infty$-induction is conservative over PA (reported in the paper [122] by Georg Kreisel).

planations in Chapter 6 have shown that this assessment as an argument against nonstandard mathematics does not stand up to a more detailed analysis, regardless of the philosophical position taken. The fact that the extension $^*\mathbb{R} \supset \mathbb{R}$ is not canonical is not a disadvantage, but rather an advantage, since extensions with additional properties (for example, a certain saturation) can be chosen if required. Concretely defined nonstandard numbers are not used in nonstandard analysis. In the nonstandard definitions of continuity, derivative, sequence limits, and so on, only the qualitative determinacy of the nonstandard numbers used is relevant (for example, that they are infinitesimal or infinitely large). The axiom of choice is used in ZFC for the construction of hyperreal numbers. However, there are conservative extensions of ZF that imply the existence of nonstandard numbers (including infinitesimals) in $\mathbb{R}$. Hence, the axiom of choice is not mandatory for nonstandard analysis (see Section 5.4.10). Nonstandard analysis is therefore just as reliable as standard analysis, foundationally speaking, and does not necessarily require stronger preconditions than standard analysis (see Section 6.2.2).

With regard to their application in the empirical sciences, both the real and the hyperreal numbers are idealizations because their construction requires the use of infinite sets that have no equivalent in our experience (see Sections 6.3.1 and 6.3.2). In addition to infinite sets, nonstandard analysis also provides infinitely large and infinitely small numbers for idealization. If such numbers appear explicitly in models of real world phenomena, they are only determined qualitatively (for example, as $x \approx 0$ or $n \gg 1$), but this does not diminish their usefulness in modeling (see Section 6.3.3). As internal set theory shows, infinitesimal and unlimited numbers exist even in $\mathbb{R}$ (if ZFC is extended to IST). Since this extension is conservative, it is epistemologically unobjectionable.

In Section 6.1.2, we examined the question of the existence of nonstandard numbers from various philosophical positions. Most mathematicians today accept ZFC as the basis for mathematics, whether as realists or formalists. Nonstandard numbers exist for ZFC or multiverse realists as hyperreal numbers or as real nonstandard numbers. Their existence cannot be ruled out in ZF realism or empirically grounded realism either. At the very least, nonstandard numbers can be assumed to be useful fictions, as Leibniz already recommended. For formalists, the question of metaphysical existence does not arise. Nonstandard numbers exist (like the real numbers) within the framework of suitable formal theories that are assumed to be consistent (and whose consistency relative to ZF is ensured). Constructivists reject infinite sets. Therefore, neither the real nor the hyperreal numbers exist for them (at least not in the way they are usually constructed). On the other hand, there are constructivist variants of both standard and nonstandard analysis.

Irrespective of the philosophical positions discussed on the foundations of mathematics, a categorical distinction between standard numbers and nonstandard numbers cannot be maintained with regard to ontology, epistemology and applicability.

7.2 What is the bottom line?

7.2.1 Options for teaching

Keisler's experimental calculus courses and the more recent teaching experiences of Katz and Ely show that a nonstandard introduction to calculus with a subsequent treatment of Weierstrass's limit concept is possible and can have positive effects on the motivation and learning success of students (see Sections 4.3.1 and 4.3.3). However, such an approach means a profound change to the established lecture concepts, and the willingness to experiment with the lectures is rather low.

In the survey on the assessment of the possible uses of nonstandard analysis in teaching, the feedback from lecturers was only occasionally positive with regard to its use in basic lectures, but almost a third were positive with regard to its use in complementary courses (see the summary in Section 7.1.2). This proportion is higher than the actual range of nonstandard courses would suggest (see Section 7.1.1).

The question therefore arises as to how this willingness can be utilized and how the potential of nonstandard analysis can be better exploited for teaching. From the considerations so far, at least three scenarios emerge as to how elements of nonstandard analysis could be integrated into teaching without extensively redesigning the basic lectures.[4]

Integration into the basic lecture (constructive approach with Fréchet filter)
With minor additions to the standard procedure, the suggestive power of nonstandard notation can be used in the way Laugwitz has already practiced and as suggested by Henle in [89] (see Section 4.2.1). By using the Fréchet filter, the direct reference to standard analysis is maintained, but the effect is nevertheless enormous, especially for the concepts of sequence limit, function limit, continuity, uniform continuity and derivative. The absorption of quantifiers in the notations $x \approx 0$ and $n \gg 1$ creates a cognitive bridge from intuition to definition or, in Tall's terms, between the *concept image* and the *concept definition*.

Complementary course (axiomatic approach with elementary extension principle)
In a complementary course (e. g., an undergraduate seminar), analysis can be built up once again using the elementary extension principle with hyperreal numbers. References to the basic lecture and considerations on the equivalence of both approaches offer the opportunity to deal intensively with the standard definitions once again and at the same time to develop an intuitive image of these concepts. The construction of

---

[4] The possibility of offering additional in-depth lectures and seminars on nonstandard analysis for intermediate and advanced semesters (for example, based on Goldblatt's lectures [85]) remains unaffected.

the hyperreal numbers, or at least a hint of it, could be discussed at the end of such a course, after the extension $^*\mathbb{R} \supset \mathbb{R}$ has been sufficiently motivated by its application in calculus.

Complementary course (axiomatic approach with language extension)
A third option is to place greater emphasis on the historical and philosophical aspects and to focus on raising awareness of the foundations of mathematics by means of a complementary course (e. g., an undergraduate seminar), as outlined in Chapter 2. The repetition of material from the basic lectures from a new perspective is a positive side effect. Further objectives of such a course were stated at the beginning of Chapter 2.

7.2.2 Closing words

In the last section, various scenarios were presented as to how elements of nonstandard analysis could be used in teaching. Practical experience and investigations of the effect from the teachers' and students' perspective would be desirable. This applies in particular to the third scenario, which was chosen as an introduction in this book (Section 1.1) and which was elaborated in Chapter 2.

Challenges posed by nonstandard analysis for familiar views of mathematics, such as the Cantor–Dedekind postulate or the standard model hypothesis, were discussed in detail in Chapter 5 and summarized again in this chapter. Overall, mathematical, philosophical, and didactic arguments against nonstandard were analyzed, reflected upon and evaluated with the result that they are often based on routines and preconceived ideas. I hope that this book has provided food for thought for an unprejudiced discourse on nonstandard analysis and its possible applications in teaching.

Bibliography

[1] Stephen Abbott. *Understanding Analysis*. Springer, 2015.
[2] Sergio Albeverio et al. *Nonstandard Methods in Stochastic Analysis and Mathematical Physics*. Vol. 122. Pure and Applied Mathematics. Orlando u. a.: Academic Press, 1986.
[3] Robert Anderson. "Infinitesimal Methods in Mathematical Economics". In: *SSRN* (2008). URL: https://dx.doi.org/10.2139/ssrn.3769003.
[4] Carolin Antos et al. "Multiverse conceptions in set theory". In: *Synthese* 192.8 (2015), pp. 2463–2488. URL: https://doi.org/10.1007/s11229-015-0819-9.
[5] Tom M. Apostol. *Calculus, Volume 1*. John Wiley & Sons, 1991.
[6] Tom Archibald et al. "A question of fundamental methodology: reply to Mikhail Katz and his coauthors". In: *The Mathematical Intelligencer* (2022), pp. 1–4.
[7] Leif O. Arkeryd, Nigel J. Cutland, and C. Ward Henson. *Nonstandard Analysis: Theory and Applications*. Vol. 493. Mathematical and Physical Sciences. Dordrecht: Springer Science+Business Media, 1997. URL: https://doi.org/10.1017/CBO9781139172110.
[8] Richard Arthur and David Rabouin. "On the unviability of interpreting Leibniz's infinitesimals through non-standard analysis". In: *Historia Mathematica* 66 (2024), pp. 26–42. DOI: https://doi.org/https://doi.org/10.1016/j.hm.2023.12.001. URL: https://www.sciencedirect.com/science/article/pii/S031508602300085X.
[9] Richard T. W. Arthur. "Leery bedfellows: Newton and Leibniz on the status of infinitesimals". In: *Infinitesimal Differences: Controversies Between Leibniz and His Contemporaries* ([86]). 2008.
[10] Jeremy Avigad. "Weak theories of nonstandard arithmetic and analysis". In: *Reverse Mathematics 2001* ([199]). 2001, pp. 19–46.
[11] Steve Awodey. "From sets to types, to categories, to sets". In: *Foundational Theories of Classical and Constructive Mathematics*. Vol. 76. Western Ontario Series in Philosophy of Sience. Dordrecht, London: Springer, 2011, pp. 113–126.
[12] Paul Bachmann. *Vorlesungen über die Natur der Irrationalzahlen*. Leipzig: Teubner, 1892.
[13] Jacques Bair et al. "Cauchy's work on integral geometry, centers of curvature, and other applications of infinitesimals". In: *Real Analysis Exchange* 45.1 (2020), pp. 127–150. DOI: https://doi.org/10.14321/realanalexch.45.1.0127. URL: https://doi.org/10.14321/realanalexch.45.1.0127.
[14] Jacques Bair et al. "Historical infinitesimalists and modern historiography of infinitesimals". In: *Antiquitates Mathematicae* (2022), pp. 189–257. URL: https://doi.org/10.14708/am.v16i1.7169.
[15] Jacques Bair et al. "Is pluralism in the history of mathematics possible?" In: *The Mathematical Intelligencer* 45.8 (2023). URL: https://doi.org/10.1007/s00283-022-10248-0.
[16] Jacques Bair et al. "Leibniz's well-founded fictions and their interpretations". In: *Matematychni Studii* 49.2 (2018), pp. 186–224. URL: https://doi.org/10.15330/ms.49.2.186-224.
[17] Jacques Bair et al. "Procedures of Leibnizian infinitesimal calculus: an account in three modern frameworks". In: *British Journal for the History of Mathematics* (2021), pp. 1–40. URL: https://doi.org/10.1080/26375451.2020.1851120.
[18] R. G. Bartle and D. R. Sherbert. *Introduction to Real Analysis*. 4th ed. John Wiley & Sons, Incorporated, 2011.
[19] Jon Barwise. *Admissible Sets and Structures: An Approach to Definability Theory*. Berlin, Heidelberg: Springer, 1975.
[20] Tiziana Bascelli et al. "Fermat, Leibniz, Euler, and the Gang: the true history of the concepts of limit and shadow". In: *Notices of the American Mathematical Society* 61.8 (2014), pp. 848–864.
[21] Heinz Bauer. *Wahrscheinlichkeitstheorie*. 5th ed. Berlin, New York: Walter de Gruyter, 2002.
[22] Ludwig Bauer. "Mathematik, Intuition, Formalisierung: eine Untersuchung von Schülerinnen- und Schülervorstellungen zu $0,\overline{9}$". In: *Journal für Mathematik-Didaktik* 32.1 (2011), pp. 79–102. URL: https://link.springer.com/content/pdf/10.1007/s13138-010-0024-9.pdf.

[23] Thomas Bedürftig. "Über die Grundproblematik der Grenzwerte". In: *Mathematische Semesterberichte* 65.2 (2018), pp. 277–298. URL: https://doi.org/10.1007/s00591-018-0220-0.

[24] Thomas Bedürftig and Karl Kuhlemann. *Grenzwerte oder infinitesimale Zahlen?* Springer Spektrum, 2020. URL: https://link.springer.com/content/pdf/10.1007/978-3-658-31908-3.pdf.

[25] Thomas Bedürftig and Roman Murawski. *Philosophie der Mathematik*. 4th ed. Berlin, Boston: de Gruyter, 2019.

[26] Thomas Bedürftig and Roman Murawski. *Philosophy of Mathematics*. 1st ed. Berlin Boston: de Gruyter, 2018.

[27] Thomas Bedürftig and Roman Murawski. *Zählen. Grundlage der elementaren Arithmetik*. Vol. 7. Studium und Lehre Mathematik. Hildesheim, Berlin: Franzbecker, 2001.

[28] Ehrhard Behrends. *Analysis Band 1. Ein Lernbuch für den sanften Wechsel von der Schule zur Uni. Von Studenten mitentwickelt*. 6th ed. Wiesbaden: Springer Spektrum, 2015.

[29] John Lane Bell. *A Primer of Infinitesimal Analysis*. 2nd ed. Cambridge University Press, 2008.

[30] John Lane Bell and Moshé Machover. *A Course in Mathematical Logic*. Amsterdam u. a.: North-Holland Publ. Comp., 1977.

[31] Benno van den Berg, Eyvind Briseid, and Pavol Safarik. "A functional interpretation for nonstandard arithmetic". In: *Annals of Pure and Applied Logic* 163.12 (2012), pp. 1962–1994. URL: https://doi.org/10.1016/j.apal.2012.07.003.

[32] George Berkeley. *The analyst; or, a discourse addressed to an infidel mathematician. Wherein it is examined whether the object, Principles, and inferences of the modern analysis are more distinctly conceived, or more evidently deduced, than religious mysteries and Points of Faith. By the author of The minute philosopher*. London: printed for J. Tonson in the Strand, 1734.

[33] Allen R. Bernstein and Abraham Robinson. "Solution of an invariant subspace problem of K. T. Smith and P. R. Halmos". In: *Pacific Journal of Mathematics* 16.3 (1966), pp. 421–431.

[34] Martin Berz. "Calculus and numerics on Levi-Civita fields". In: *Computational Differentiation: Techniques, Applications, and Tools*. 1996, pp. 19–37.

[35] Jan Bezuidenhout. "Limits and continuity: some conceptions of first-year students". In: *International Journal of Mathematical Education in Science and Technology* 32.4 (2001), pp. 487–500. URL: https://doi.org/10.1080/00207390010022590.

[36] Errett Bishop. "Book review: elementary calculus". In: *Bulletin of the American Mathematical Society* 83.2 (1977), pp. 205–208.

[37] Errett Bishop. *Foundations of Constructive Analysis*. Vol. 60. McGraw-Hill Series in Higher Mathematics. New York u. a.: McGraw-Hill, 1967.

[38] Errett Bishop. "The crisis in contemporary mathematics". In: *Historia Mathematica* 2.4 (1975), pp. 507–517.

[39] Errett Bishop and Douglas Bridges. *Constructive Analysis*. Vol. 279. Grundlehren der Mathematischen Wissenschaften, A Series of Comprehensive Studies in Mathematics. Berlin, Heidelberg: Springer, 1985. URL: https://doi.org/10.1007/978-3-642-61667-9.

[40] Emanuele Bottazzi and Mikhail G. Katz. "Infinite lotteries, spinners, applicability of hyperreals". In: *Philosophia Mathematica* 29.1 (Oct. 2020), pp. 88–109. URL: https://doi.org/10.1093/philmat/nkaa032.

[41] Emanuele Bottazzi and Mikhail G. Katz. "Internality, transfer, and infinitesimal modeling of infinite processes". In: *Philosophia Mathematica* 29.2 (Sept. 2020), pp. 256–277. URL: https://doi.org/10.1093/philmat/nkaa033.

[42] Carl B. Boyer. *The History of the Calculus and Its Conceptual Development: (The Concepts of the Calculus)*. Courier Corporation, 1959.

[43] David Bressoud et al. *Teaching and Learning of Calculus*. Cham: Springer Nature, 2016. URL: https://doi.org/10.1007/978-3-319-32975-8.

[44] Georg Cantor. *Gesammelte Abhandlungen mathematischen und philosophischen Inhalts*. Ed. by Ernst Zermelo. Berlin, Heidelberg: Springer, 1932. URL: https://doi.org/10.1007/978-3-662-00274-2.

[45] Georg Cantor. "Über die Ausdehnung eines Satzes aus der Theorie der trigonometrischen Reihen". In: *Mathematische Annalen* 5.1 (1872), pp. 123–132. URL: https://doi.org/10.1007/BF01446327.
[46] Augustin-Louis Cauchy. *Théorie de la propagation des ondes à la surface d'un fluide pesant d'une profondeur indéfinie (published 1827, with additional Notes)*. Oeuvres complètes, Series 1, Vol. 1. Académie royale des sciences, 1815, pp. 4–318.
[47] Chen Chung Chang and H. Jerome Keisler. *Model Theory*. 3rd ed. Vol. 73. Studies in Logic and the Foundations of Mathematics. Amsterdam, New York: North-Holland Press, 1990.
[48] Chitat Chong et al. eds. *Infinity and Truth*. Vol. 25. Lecture Notes Series, Institute for Mathematical Sciences, National University of Singapore. World Scientific, 2014.
[49] Paul J. Cohen. "Comments on the foundations of set theory". In: *Axiomatic Set Theory, Part 1* ([193]). 1971, pp. 9–15.
[50] Paul J. Cohen. *Set Theory and the Continuum Hypothesis*. Reading, Mass., New York u. a.: Benjamin, 1966.
[51] Paul J. Cohen. "The independence of the continuum hypothesis". In: *Proceedings of the National Academy of Sciences of the United States of America* 50.1 (1963), pp. 1143–1148.
[52] Paul J. Cohen. "The independence of the continuum hypothesis, II". In: *Proceedings of the National Academy of Sciences of the United States of America* 51.1 (1964), pp. 105–110.
[53] Alain Connes. "Cyclic Cohomology, Noncommutative Geometry and Quantum Group Symmetries". In: Dec. 2003, pp. 1–71. URL: https://doi.org/10.1007/978-3-540-39702-1_1.
[54] Bernard Cornu. "Limits". In: *Advanced Mathematical Thinking* ([214]). 1991, pp. 153–166.
[55] Daniel W. Cunningham. *Set Theory*. eng. Cambridge Mathematical Textbooks. New York, NY: Cambridge University Press, 2016. ISBN: 9781107120327.
[56] Nigel Cutland. *Nonstandard Analysis and Its Applications*. Vol. 10. London Mathematical Society Student Texts. Cambridge University Press, 1988. URL: https://doi.org/10.1017/CBO9781139172110.
[57] Martin Davis. *Applied Nonstandard Analysis*. New York u. a.: Wiley, 1977.
[58] Philip J. Davis, Reuben Hersh, and Elena Anne Marchisotto. *The Mathematical Experience*, study ed. Springer, 2012.
[59] Richard Dedekind. *Was sind und was sollen die Zahlen? Stetigkeit und Irrationale Zahlen*. Klassische Texte der Wissenschaft. Berlin: Springer Spektrum, 2017. URL: https://doi.org/10.1007/978-3-662-54339-9.
[60] André Deledicq and Marc Diener. *Leçons de calcul infinitésimal*. Paris: Armand Colin, 1989.
[61] Francine Diener and Marc Diener, eds. *Nonstandard Analysis in Practice*. Berlin, Heidelberg: Springer, 1995.
[62] Bruno Dinis and Imme van den Berg. *Neutrices and External Numbers: A Flexible Number System*. Boca Raton: Chapman and Hall/CRC, 2019. URL: https://doi.org/10.1201/9780429291456.
[63] Kenny Easwaran. "Regularity and hyperreal credences". In: *Philosophical Review* 123.1 (2014), pp. 1–41. URL: https://www.jstor.org/stable/44282331.
[64] Heinz-Dieter Ebbinghaus. *Einführung in die Mengenlehre*. 5th ed. Berlin, Heidelberg: Springer Spektrum, 2021. URL: https://doi.org/10.1007/978-3-662-63866-8.
[65] Heinz-Dieter Ebbinghaus et al. *Mathematical Logic*. Vol. 1910. Springer, 1994.
[66] Heinz-Dieter Ebbinghaus et al. *Zahlen*. 3rd ed. Berlin, Heidelberg: Springer, 1992. URL: https://doi.org/10.1007/978-3-642-58155-7.
[67] C. H. Edwards jr. *The Historical Development of the Calculus*. Springer Science & Business Media, 2012.
[68] Philip Ehrlich. "The absolute arithmetic continuum and the unification of all numbers great and small". In: *Bulletin of Symbolic Logic* 18.1 (2012), pp. 1–45. URL: https://doi.org/10.2178/bsl/1327328438.
[69] Philip Ehrlich. "The rise of non-Archimedean mathematics and the roots of a misconception I: the emergence of non-Archimedean systems of magnitudes". In: *Archive for History of Exact Sciences* 60 (2006), pp. 1–121. URL: https://doi.org/10.1007/s00407-005-0102-4.
[70] Robert Ely. "Nonstandard student conceptions about infinitesimals". In: *Journal for Research in Mathematics Education* 41.2 (2010), pp. 117–146. URL: https://www.jstor.org/stable/20720128.

[71] Robert Ely. *Student Obstacles and Historical Obstacles to Foundational Concepts of Calculus*. University of Wisconsin–Madison, 2007. URL: https://hdl.handle.net/2027/wu.89097475107.

[72] Robert Ely. "Teaching calculus with infinitesimals and differentials". In: *ZDM Mathematics Education* 53 (2020), pp. 591–604. URL: https://doi.org/10.1007/s11858-020-01194-2.

[73] Paul Erdős, Leonard Gillman, and Melvin Henriksen. "An isomorphism theorem for real-closed fields". In: *Annals of Mathematics* 61.3 (1955), pp. 542–554. URL: https://doi.org/10.2307/1969812.

[74] Euklid. *Die Elemente: Buch I-XIII. Nach Heibergs Text aus dem Griechischen übersetzt und herausgegeben von Clemens Thaer*. Darmstadt: Wissenschaftliche Buchgesellschaft, 1975.

[75] Peter Fletcher et al. "Approaches to analysis with infinitesimals following Robinson, Nelson, and others". In: *Real Analysis Exchange* 42.2 (2017), pp. 193–252. URL: https://doi.org/10.14321/realanalexch.42.2.0193.

[76] Abraham Adolf Fraenkel. *Einleitung in die Mengenlehre*. 3rd ed. Vol. 9. Die Grundlehren der Mathematischen Wissenschaften in Einzeldarstellungen mit besonderer Berücksichtigung der Anwendungsgebiete. Berlin, Heidelberg: Springer, 1928. URL: https://doi.org/10.1007/978-3-662-42029-4.

[77] Harvey Friedman. "Some systems of second order arithmetic and their use". In: *Proceedings of the International Congress of Mathematicians (Vancouver, BC, 1974)*. Vol. 1. Citeseer. 1975, pp. 235–242.

[78] Carl Immanuel Gerhardt, ed. *Die philosophischen Schriften von Gottfried Wilhelm Leibniz*. Berlin: Weidmann, 1875–1890.

[79] Victoria Gitman and Joel David Hamkins. "A natural model of the multiverse axioms". In: *Notre Dame Journal of Formal Logic* 51.4 (2010), pp. 475–484. URL: https://doi.org/10.1215/00294527-2010-030.

[80] Kurt Gödel. "Consistency-proof for the generalized continuum-hypothesis". In: *Proceedings of the National Academy of Sciences* 25 (1939), pp. 220–224.

[81] Kurt Gödel. "Russell's mathematical logic". In: *The Philosophy of Bertrand Russell*. Ed. by Paul Arthur Schilpp. Evanstown: Northwestern University, 1944, pp. 123–153.

[82] Kurt Gödel. "The consistency of the axiom of choice and the generalized continuum hypothesis". In: *Proceedings of the National Academy of Sciences* 24 (1938), pp. 556–557.

[83] Kurt Gödel. *The Consistency of the Continuum Hypothesis*. Annals of Mathematics Studies No. 3. Princeton, N. J.: Princeton University Press, 1940.

[84] Kurt Gödel. "Über eine bisher noch nicht benützte Erweiterung des finiten Standpunktes". In: *Dialectica* 12.3–4 (1958), pp. 280–287.

[85] Robert Goldblatt. *Lectures on the Hyperreals: An Introduction to Nonstandard Analysis*. Vol. 188. Graduate Texts in Mathematics. New York, NY: Springer, 1998. URL: https://doi.org/10.1007/978-1-4612-0615-6.

[86] Ursula Goldenbaum and Douglas Jesseph, eds. *Infinitesimal Differences: Controversies Between Leibniz and His Contemporaries*. Berlin, New York: Walter de Gruyter, 2008.

[87] Joel David Hamkins. "Are we correct in thinking we have an absolute concept of the finite?" In: *Infinity and Truth* ([48]). 2014, pp. 230–231.

[88] Joel David Hamkins. "The set-theoretic multiverse". In: *The Review of Symbolic Logic* 5.3 (2012), pp. 416–449. URL: https://doi.org/10.1017/S1755020311000359.

[89] James M. Henle. "Non-nonstandard analysis: real infinitesimals". In: *The Mathematical Intelligencer* 21.1 (1999), pp. 67–73.

[90] James M. Henle and Eugene M. Kleinberg. *Infinitesimal Calculus*. Cambridge, MA: MIT Press, 1979.

[91] Luz Marina Hernandez and Jorge M. Lopez Fernandez. "Teaching calculus with infinitesimals: new perspectives". In: *The Mathematics Enthusiast* 15.3 (2018), pp. 371–390. URL: https://scholarworks.umt.edu/tme/vol15/iss3/2.

[92] Edwin Hewitt. "Rings of real-valued continuous functions. I". In: *Transactions of the American Mathematical Society* 64.1 (1948), pp. 45–99.

[93] Arend Heyting. "Die formalen Regeln der intuitionistischen Logik". In: *Sitzungsbericht Preußische Akademie der Wissenschaften Berlin, physikalisch-mathematische Klasse II* (1930), pp. 42–56.

[94] David Hilbert. "On the infinite". In: *Philosophy of Mathematics: Selected Readings*. Ed. by Paul Benacerraf and Hilary Putnam. Cambridge University Press, 1984, pp. 183–201.

[95] David Hilbert. "Über das Unendliche". In: *Mathematische Annalen* 95.1 (1926), pp. 161–190. URL: https://doi.org/10.1007/BF01206605.

[96] David Hilbert. "Über den Zahlbegriff". In: *Jahresberichte der DMV* 8 (1900).

[97] Joram Hirschfeld. "The nonstandard treatment of Hilbert's fifth problem". In: *Transactions of the American Mathematical Society* 321.1 (1990), pp. 379–400. URL: https://doi.org/10.2307/2001608.

[98] Paul Howard and Jean E. Rubin. *Consequences of the Axiom of Choice*. Vol. 59. Mathematical Surveys and Monographs. Providence, RI: American Mathematical Society, 1998.

[99] Karel Hrbaček. "Axiom of choice in nonstandard set theory". In: *Journal of Logic and Analysis* 4.8 (Apr. 2012), pp. 1–9. URL: https://doi.org/10.4115/jla.2012.4.8.

[100] Karel Hrbaček. "Axiomatic foundations for nonstandard analysis". In: *Fundamenta Mathematicae* 98 (1978), pp. 1–19.

[101] Karel Hrbaček. "Relative set theory: internal view". *Journal of Logic and Analysis* 1.8 (2009), pp. 1–108. URL: https://doi.org/10.4115/jla.2009.1.8.

[102] Karel Hrbaček. "Relative set theory: some external issues". In: *Journal of Logic and Analysis* 2.8 (2010), pp. 1–37. URL: https://doi.org/10.4115/jla.2010.2.8.

[103] Karel Hrbaček and Mikhail G. Katz. "Constructing nonstandard hulls and Loeb measures in internal set theories". In: *Bulletin of Symbolic Logic* 29.1 (2023), pp. 97–127.

[104] Karel Hrbaček and Mikhail G. Katz. "Infinitesimal analysis without the axiom of choice". In: *Annals of Pure and Applied Logic* 172.6 (2021), p. 102959. URL: https://doi.org/10.1016/j.apal.2021.102959.

[105] Karel Hrbaček, Olivier Lessmann, and Richard O'Donovan. *Analysis with Ultrasmall Numbers*. Boca Raton, London, New York: Chapman-Hall/CRC Press, 2014.

[106] Thomas J. Jech. *Set Theory*. 3rd ed. Berlin, Heidelberg: Springer, 2003. URL: https://doi.org/10.1007/3-540-44761-X.

[107] Thomas J. Jech. *The Axiom of Choice*. Vol. 75. Studies in Logic and the Foundations of Mathematics. Amsterdam u. a.: North-Holland Publ. Co., 1973.

[108] Vladimir Kanovei. "Uniqueness, collection, and external collapse of cardinals in IST and models of Peano arithmetic". In: *The Journal of Symbolic Logic* 60.1 (1995), pp. 318–324.

[109] Vladimir Kanovei, Mikhail G. Katz, and Thomas Mormann. "Tools, objects, and chimeras: Connes on the role of hyperreals in mathematics". In: *Foundations of Science* 18.2 (2013), pp. 259–296. URL: https://doi.org/10.1007/s10699-012-9316-5.

[110] Vladimir Kanovei and Michael Reeken. *Nonstandard Analysis: Axiomatically*. Berlin, Heidelberg, New York: Springer-Verlag, 2004.

[111] Vladimir Kanovei and Saharon Shelah. "A definable nonstandard model of the reals". In: *The Journal of Symbolic Logic* 69.1 (2004), pp. 159–164. URL: https://doi.org/10.2178/jsl/1080938834.

[112] Vladimir G. Kanovei. "Undecidable hypotheses in Edward Nelson's internal set theory". In: *Russian Mathematical Surveys* 46.6 (1991), pp. 1–54. URL: https://doi.org/10.1070/RM1991v046n06ABEH002870.

[113] Mikhail G. Katz, Karl Kuhlemann, and David Sherry. "A Leibniz/NSA comparison". In: *London Mathematical Society Newsletter* 511 16.12 (2024), pp. 23–27.

[114] Mikhail G. Katz and Luie Polev. "From Pythagoreans and Weierstrassians to true infinitesimal calculus". In: *Journal of Humanistic Mathematics* 7.1 (2017), pp. 87–104. URL: https://doi.org/10.5642/jhummath.201701.07.

[115] Mikhail G. Katz, David Sherry, and Monica Ugaglia. "Of Pashas, Popes, and Indivisibles". In: *Science in Context* (to appear) (2024).

[116] Mikhail G. Katz et al. "Leibniz on bodies and infinities: rerum natura and mathematical fictions". In: *The Review of Symbolic Logic* 17.1 (2024), pp. 36–66. URL: https://doi.org/10.1017/S1755020321000575.

[117] Mikhail G. Katz et al. "Two-track depictions of Leibniz's fictions". In: *The Mathematical Intelligencer* 44.3 (2022), pp. 261–266.

[118] H. Jerome Keisler. *Elementary Calculus – An Infinitesimal Approach*. 1st ed. Boston, Massachusetts: Prindle, Weber & Schmidt, 1976.
[119] H. Jerome Keisler. *Elementary Calculus – An Infinitesimal Approach*. 3rd ed. Dover Publications Inc., 2012.
[120] H. Jerome Keisler. *Elementary Calculus: An Infinitesimal Approach*. Madison: University of Wisconsin, 2012. URL: https://people.math.wisc.edu/~keisler/calc.html (visited on 08/21/2021).
[121] H. Jerome Keisler. *Foundations of Elementary Calculus*. Madison: University of Wisconsin, 2007. URL: https://people.math.wisc.edu/~keisler/foundations.html (visited on 08/21/2021).
[122] Georg Kreisel. "Axiomatizations of nonstandard analysis that are conservative extensions of formal systems for classical standard analysis". In: *Applications of Model Theory to Algebra, Analysis, and Probability* ([145]). 1969, pp. 93–108.
[123] Karl Kuhlemann. *Der Untergang von Mathemagika: ein Roman über eine Welt jenseits unserer Vorstellung*. Springer-Verlag, 2015.
[124] Karl Kuhlemann. "Nichtstandard in der elementaren Analysis. Mathematische, logische, philosophische und didaktische Studien zur Bedeutung der Nichtstandardanalysis in der Lehre". Dissertation. Gottfried Wilhelm Leibniz Universität Hannover, 2022. URL: https://doi.org/10.15488/12105.
[125] Karl Kuhlemann. "Über die Technik der infiniten Vergrößerung und ihre mathematische Rechtfertigung". In: *Siegener Beiträge zur Geschichte und Philosophie der Mathematik*. Ed. by Ralf Krömer and Gregor Nickel. Vol. 10. 2018, pp. 47–65. URL: https://nbn-resolving.org/urn:nbn:de:hbz:467-14260.
[126] Edmund Landau. *Foundations of Analysis*. 79. American Mathematical Soc., 2001.
[127] Dieter Landers and Lothar Rogge. *Nichtstandard Analysis*. Berlin Heidelberg: Springer, 1994.
[128] Serge Lang. *Undergraduate Analysis*. Springer Science & Business Media, 2013.
[129] Detlef Laugwitz. *Infinitesimalkalkül: Eine elementare Einführung in die Nichtstandard-Analysis*. Mannheim, Wien, Zürich: Bibliographisches Institut, 1978.
[130] Detlef Laugwitz. *Zahlen und Kontinuum*. Mannheim, Wien, Zürich: Bibliographisches Institut, 1986.
[131] F. William Lawvere. "Categorical dynamics". In: *Topos Theoretic Methods in Geometry*. Ed. by Anders Kock. Vol. 30. Various Publications Series. Mathematisk Institut, Aarhus Universitet, 1979, pp. 1–28.
[132] F. William Lawvere. "Toward the description in a smooth topos of the dynamically possible motions and deformations of a continuous body". In: *Cahiers de Topologie et Géométrie Différentielle Catégoriques* 21.4 (1980), pp. 377–392. URL: http://www.numdam.org/article/CTGDC_1980__21_4_377_0.pdf (visited on 08/22/2021).
[133] Gottfried Wilhelm Leibniz. *Letter to l'Hospital, 14/24 June 1695*. A.III, 6, N. 135. 1695.
[134] Gottfried Wilhelm Leibniz. *Letter to Pinsson, end of September 1701*. A.I, 20, N. 290. 1701.
[135] Gottfried Wilhelm Leibniz. *Letter to Masson*. in [78], Bd. VI, 624–629. 1716.
[136] Gottfried Wilhelm Leibniz. *Letter to Varignon, 2 February 1702*. A.III, 9, N. 5. 1702.
[137] Gottfried Wilhelm Leibniz. *Letter to Varignon, 20 June 1702*. A.III, 9, N. 35. 1702.
[138] Gottfried Wilhelm Leibniz. *De quadratura arithmetica circuli ellipseos et hyperbolae*. Ed. by Eberhard Knobloch. Berlin, Heidelberg: Springer Spektrum, 2016. URL: https://doi.org/10.1007/978-3-662-52803-7.
[139] Gottfried Wilhelm Leibniz. *Leibniz-Edition, Akademie-Ausgabe. Sämtliche Schriften und Briefe*. Zitiert als A. Reihe, Band, Nummer (zum Beispiel A.III, 9, N. 5). 1923–. URL: https://leibnizedition.de.
[140] Samuel Levey, S. Shapiro, and G. Hellman. "The continuum, the infinitely small, and the law of continuity in Leibniz". In: *Shapiro–Hellman (2021)* [196] (2021), pp. 123–157.
[141] Peter A. Loeb and Manfred P. H. Wolff. *Nonstandard Analysis for the Working Mathematician*. 2nd ed. Dordrecht u. a.: Springer, 2015. URL: https://doi.org/10.1007/978-94-017-7327-0.
[142] Paul Lorenzen. *Differential und Integral: eine konstruktive Einführung in die klassische Analysis*. Frankfurt am Main: Akademische Verlagsgesellschaft, 1965.

[143] Jerzy Łoś. "Quelques remarques, théorèmes et problèmes sur les classes définissables d'algèbres". In: *Mathematical Interpretation of Formal Systems* (1955), pp. 98–113.
[144] Robert Lutz and Michel Goze. *Nonstandard Analysis: A Practical Guide with Applications*. Vol. 881. Lecture Notes in Mathematics. Berlin, Heidelberg, New York: Springer, 1981. URL: https://doi.org/10.1007/BFb0093397.
[145] Wilhelmus A. J. Luxemburg, ed. *Applications of Model Theory to Algebra, Analysis, and Probability*. New York: Holt, Rinehart and Winston, 1969.
[146] Moshe Machover. "The place of nonstandard analysis in mathematics and in mathematics teaching". In: *The British Journal for the Philosophy of Science* 44.2 (1993), pp. 205–212. URL: https://doi.org/10.1093/bjps/44.2.205.
[147] Penelope Maddy. *Naturalism in Mathematics*. Oxford: Clarendon Press, 1997.
[148] Penelope Maddy. *Realism in Mathematics*. Oxford: Clarendon Press, 1990.
[149] Penelope Maddy et al. *Second Philosophy: A Naturalistic Method*. Oxford u. a.: Oxford University Press, 2007.
[150] Jerrold E. Marsden and Michael J. Hoffman. *Elementary Classical Analysis*. Macmillan, 1993.
[151] Philipp Mayring. "Qualitative content analysis: theoretical foundation, basic procedures and software solution". In: *SSOAR Open Access Repository* (2014). URL: https://nbn-resolving.org/urn:nbn:de:0168-ssoar-395173.
[152] Philipp Mayring. *Qualitative Inhaltsanalyse*. 12th ed. Weinheim, Basel: Beltz Verlag, 2015.
[153] Philipp Mayring et al. "Qualitative content analysis". In: *A Companion to Qualitative Research* 1.2 (2004), pp. 159–176.
[154] Herbert Meschkowski. "Aus den Briefbüchern Georg Cantors". In: *Archive for History of Exact Sciences* 2.6 (1965), pp. 503–519. URL: https://doi.org/10.1007/BF00324881.
[155] John Stuart Mill. *A System of Logic, Ratiocinative and Inductive: Being a Connected View of the Principles of Evidence and the Methods of Scientific Investigation*. London: Longmans, Green, Reader, and Dyer, 1875. URL: https://nl.sub.uni-goettingen.de/volumes/id/F101006848.
[156] James Donald Monk. *Mathematical Logic*. Vol. 37. Graduate Texts in Mathematics. New York, Heidelberg, Berlin: Springer-Verlag, 1976. URL: https://doi.org/10.1007/978-1-4684-9452-5.
[157] Gregory H. Moore. *Zermelo's Axiom of Choice: Its Origins, Development, and Influence*. New York, Heidelberg, Berlin: Springer-Verlag, 1982. URL: https://doi.org/10.1007/978-1-4613-9478-5.
[158] Roman Murawski. "On the philosophical meaning of reverse mathematics". In: *Philosophie der Mathematik: Akten des 15. Internationalen Wittgenstein-Symposiums: 16. bis 23. August 1992, Kirchberg am Wechsel (Österreich)*. Vol. 1. Wien: Hölder-Pichler-Tempsky, 1993, pp. 173–184.
[159] Edward Nelson. *Hilbert's Mistake*. 2007. URL: www.math.princeton.edu/~nelson/papers/hm.pdf (visited on 08/25/2021).
[160] Edward Nelson. "Internal set theory: a new approach to nonstandard analysis". In: *Bulletin of the American Mathematical Society* 83.6 (1977), pp. 1165–1198.
[161] Edward Nelson. *Predicative Arithmetic*. Vol. 32. Mathematical Notes. Princeton, NJ u. a.: Princeton University Press, 1986.
[162] Edward Nelson. *Radically Elementary Probability Theory, Annals of Mathematics Studies*. Vol. 117. Annals of Mathematics Studies. Princeton, NJ u. a.: Princeton University Press, 1987. URL: https://doi.org/10.1515/9781400882144.
[163] Michael Oehrtman. "Collapsing dimensions, physical limitation, and other student metaphors for limit concepts". In: *Journal for Research in Mathematics Education* 40.4 (2009), pp. 396–426. URL: https://www.jstor.org/stable/40539345.
[164] Michael Oehrtman, Craig Swinyard, and Jason Martin. "Problems and solutions in students' reinvention of a definition for sequence convergence". In: *The Journal of Mathematical Behavior* 33 (2014), pp. 131–148. URL: https://doi.org/10.1016/j.jmathb.2013.11.006.
[165] Erik Palmgren. "A sheaf-theoretic foundation for nonstandard analysis". In: *Annals of Pure and Applied Logic* 85.1 (1997), pp. 69–86.

[166] Erik Palmgren. "Developments in constructive nonstandard analysis". In: *Bulletin of Symbolic Logic* (1998), pp. 233–272. URL: https://www.jstor.org/stable/421031.
[167] Matthew W. Parker. "Symmetry arguments against regular probability: a reply to recent objections". In: *European Journal for Philosophy of Science* 9.1 (2019), pp. 1–21.
[168] Giuseppe Peano. *Opere scelte: Analisi matematica. Calcolo numerico*. Vol. 1. Edizioni cremonese, 1957.
[169] Giuseppe Peano. "Sugli ordini degli infiniti". In: *Rendiconti della R. Accademia dei Lincei* 19 (1910), pp. 778–781.
[170] Yves Péraire. "Théorie relative des ensembles internes". In: *Osaka Journal of Mathematics* 29.2 (1992), pp. 267–297.
[171] David Pincus and Robert M. Solovay. "Definability of measures and ultrafilters". In: *The Journal of Symbolic Logic* 42.2 (1977), pp. 179–190.
[172] Siegmund Probst. "Indivisibles and infinitesimals in early mathematical texts of Leibniz". In: *Infinitesimal Differences: Controversies Between Leibniz and His Contemporaries* ([86]). Ed. by Ursula Goldenbaum and Douglas Jesseph. Berlin, New York: Walter de Gruyter, 2008, pp. 95–106. URL: https://doi.org/10.1515/9783110211863.95.
[173] Alexander R. Pruss. "Infinitesimals are too small for countably infinite fair lotteries". In: *Synthese* 191.6 (2014), pp. 1051–1057. URL: https://doi.org/10.1007/s11229-013-0307-z.
[174] Alexander R. Pruss. "Underdetermination of infinitesimal probabilities". In: *Synthese* 198.1 (2021), pp. 777–799. URL: https://doi.org/10.1007/s11229-018-02064-x.
[175] Walter Purkert. "Infinitesimalrechnung für Ingenieure—Kontroversen im 19. Jahrhundert". In: *Rechnen mit dem Unendlichen* ([208]). Springer, 1990, pp. 179–192.
[176] Hilary Putnam. "What is mathematical truth?" In: *Historia Mathematica* 2.4 (1975), pp. 529–533. URL: https://doi.org/10.1016/0315-0860(75)90116-0.
[177] David Rabouin and Richard T. W. Arthur. "Leibniz's syncategorematic infinitesimals II: their existence, their use and their role in the justification of the differential calculus". In: *Archive for History of Exact Sciences* 74 (2020), pp. 401–443. URL: https://doi.org/10.1007/s00407-020-00249-w.
[178] Alain M. Robert. *Nonstandard Analysis*. New York u. a.: John Wiley & Sons, 1988.
[179] Abraham Robinson. "Concerning progress in the philosophy of mathematics". In: *Studies in Logic and the Foundations of Mathematics*. Vol. 80. Elsevier, 1975, pp. 41–52.
[180] Abraham Robinson. "Formalism 64". In: *Logic, Methodology and Philosophy of Science: Proceedings of the 1964 International Congress, held at the Hebrew University of Jerusalem, Israel, from August 26 to September 2, 1964*. Amsterdam u. a.: North-Holland, 1965, pp. 228–246.
[181] Abraham Robinson. "Non-standard analysis". In: *Indagationes Mathematicae* 23 (1961), pp. 432–440.
[182] Abraham Robinson. *Non-Standard Analysis*. Studies in Logic and the Foundations of Mathematics. Amsterdam: North-Holland Pub. Co., 1966.
[183] Abraham Robinson. *Non-Standard Analysis*. 2nd ed. Studies in Logic and the Foundations of Mathematics. Amsterdam: North-Holland Pub. Co., 1974.
[184] Abraham Robinson. *Selected Papers*. North Holland, 1979.
[185] Abraham Robinson. "Some thoughts on the history of mathematics". In: *Compositio Mathematica*, Vol. 20 (1968), pp. 188–193. URL: http://www.numdam.org/item/CM_1968__20__188_0/.
[186] Abraham Robinson and Elias Zakon. "A set-theoretical characterization of enlargements". In: *Applications of Model Theory to Algebra, Analysis and Probability (Proceedings of the 1967 International Symposium, California Institute of Technology)*. Ed. by Wilhelmus Anthonius Josephus Luxemburg. New York u. a.: Holt, Rinehart and Winston, 1969, pp. 109–122.
[187] H. L. Royden and P. M. Fitzpatrick. *Real Analysis*. 4th ed. Prentice Hall, 2010.
[188] Walter Rudin. *Principles of Mathematical Analysis*. 3rd ed. New York u. a.: McGraw-Hill, 1976.
[189] Sam Sanders. "The unreasonable effectiveness of nonstandard analysis". In: *Journal of Logic and Computation* 30.1 (2020), pp. 459–524. URL: https://doi.org/10.1093/logcom/exaa019.
[190] Curt Schmieden and Detlef Laugwitz. "Eine Erweiterung der Infinitesimalrechnung". In: *Mathematische Zeitschrift* 69 (1958), pp. 1–39. URL: https://doi.org/10.1007/BF01187391.

[191] Walter Schnitzspan. "Konstruktive Nonstandard-Analysis". PhD thesis. Technische Hochschule Darmstadt, 1976.
[192] Erhard Scholz. "Herman Weyl on the concept of continuum". In: *Proof Theory: History and Philosophical Significance*. Springer, 2000, pp. 195–217.
[193] Dana S. Scott and Thomas J. Jech, eds. *Axiomatic Set Theory, Part 1*. Vol. 13, 1. Proceedings of Symposia in Pure Mathematics. Providence, RI: American Mathematical Society, 1971.
[194] John Selden and Annie Selden. "Unpacking the logic of mathematical statements". In: *Educational Studies in Mathematics* 29.2 (1995), pp. 123–151.
[195] Stewart Shapiro. *Philosophy of Mathematics: Structure and Ontology*. Oxford u. a.: Oxford University Press, 1997.
[196] Stewart Shapiro and Geoffrey Hellman. *The History of Continua: Philosophical and Mathematical Perspectives*. Oxford University Press, USA, 2021.
[197] Joseph R. Shoenfield. *Mathematical Logic*. Boca Raton, London, New York: CRC Press, 1967. URL: https://doi.org/10.1201/9780203749456.
[198] Stephen G. Simpson. *Potential Versus Actual Infinity: Insights from Reverse Mathematics*. Annual Logic Lecture, Group in Philosophical and Mathematical Logic, University of Connecticut, April 1–3, 2015. 2015. URL: http://www.personal.psu.edu/t20/talks/uconn1504/talk.pdf (visited on 08/28/2021).
[199] Stephen G. Simpson, ed. *Reverse Mathematics 2001*. Vol. 21. Lecture Notes in Logic. Cambridge: Cambridge University Press, 2016. URL: https://doi.org/10.1017/9781316755846.
[200] Stephen G. Simpson. *Subsystems of Second Order Arithmetic*. 2nd ed. Cambridge: Cambridge University Press, 2009. URL: https://doi.org/10.1017/CBO9780511581007.
[201] Stephen G. Simpson. "Toward objectivity in mathematics". In: *Infinity and Truth* ([48]). World Scientific, 2014, pp. 157–169.
[202] Thoralf Skolem. *Begründung der elementaren Arithmetik durch die rekurrierende Denkweise ohne Anwendung scheinbarer Veränderlichen mit unendlichem Ausdehnungsbereich*. Videnskapsselskapets Skrifter. 1. Mat.-Naturv. Klasse. 1923. No. 6. Kristiania: Jacob Dybwad, 1923.
[203] Thoralf Skolem. "Über die Nicht-charakterisierbarkeit der Zahlenreihe mittels endlich oder abzählbar unendlich vieler Aussagen mit ausschliesslich Zahlenvariablen". In: *Fundamenta Mathematicae* 23.1 (1934), pp. 150–161.
[204] Robert M. Solovay. "A model of set-theory in which every set of reals is Lebesgue measurable". In: *Annals of Mathematics* 92.1 (1970), pp. 1–56. URL: https://www.jstor.org/stable/1970696.
[205] Thomas Sonar. *3000 Years of Analysis: Mathematics in History and Culture*. Springer Nature, 2020.
[206] Detlef D. Spalt. *A Brief History of Analysis: With Emphasis on Philosophy, Concepts, and Numbers, Including Weierstraß'Real Numbers*. Springer Nature, 2022.
[207] Detlef D. Spalt. *Die Analysis im Wandel und Widerstreit*. Freiburg, München: Karl Alber, 2015.
[208] Detlef D. Spalt. *Rechnen mit dem Unendlichen: Beiträge zur Entwicklung eines kontroversen Gegenstandes*. Basel, Boston Berlin: Birkhäuser, 1990. URL: https://doi.org/10.1007/978-3-0348-5242-5.
[209] Michael Spivak. *Calculus*. Cambridge University Press, 2006.
[210] Kathleen Sullivan. "The teaching of elementary calculus using the nonstandard analysis approach". In: *The American Mathematical Monthly* 83.5 (1976), pp. 370–375. URL: https://doi.org/10.1080/00029890.1976.11994130.
[211] Craig Swinyard. "Reinventing the formal definition of limit: the case of Amy and Mike". In: *The Journal of Mathematical Behavior* 30.2 (2011), pp. 93–114. URL: https://doi.org/10.1016/j.jmathb.2011.01.001.
[212] Craig Swinyard and Sean Larsen. "Coming to understand the formal definition of limit: insights gained from engaging students in reinvention". In: *Journal for Research in Mathematics Education* 43.4 (2012), pp. 465–493. URL: https://doi.org/10.5951/jresematheduc.43.4.0465.
[213] William W. Tait. "Finitism". In: *The Journal of Philosophy* 78.9 (1981), pp. 524–546. URL: https://www.jstor.org/stable/2026089.

[214] David O. Tall, ed. *Advanced Mathematical Thinking*. Vol. 11. Mathematics Education Library. New York, Dordrecht, London: Kluwer Academic Publishers, 1991.
[215] David O. Tall. "Inconsistencies in the learning of calculus and analysis". In: *Focus on Learning Problems in Mathematics* 12.3 & 4 (1990), pp. 49–63. URL: http://homepages.warwick.ac.uk/staff/David.Tall/pdfs/dot1990b-inconsist-focus.pdf (visited on 08/29/2021).
[216] David O. Tall. *Intuitive infinitesimals in the calculus*. Poster presented at the Fourth International Congress on Mathematical Education, Berkeley, 1980, with abstract appearing in Abstracts of short communications, page C5. 1980. URL: http://homepages.warwick.ac.uk/staff/David.Tall/pdfs/dot1980c-intuitive-infls.pdf (visited on 08/29/2021).
[217] David O. Tall. "Looking at graphs through infinitesimal microscopes, windows and telescopes". In: *The Mathematical Gazette* 64.427 (1980), pp. 22–49. URL: https://www.jstor.org/stable/3615886.
[218] David O. Tall and Shlomo Vinner. "Concept image and concept definition in mathematics with particular reference to limits and continuity". In: *Educational Studies in Mathematics* 12.2 (1981), pp. 151–169. URL: https://doi.org/10.1007/BF00305619.
[219] Terence Tao. *A cheap version of nonstandard analysis*. 2012. URL: https://terrytao.wordpress.com/2012/04/02/a-cheap-version-of-nonstandard-analysis/ (visited on 04/03/2021).
[220] Terence Tao. *Hilbert's Fifth Problem and Related Topics*. Vol. 153. Graduate Studies in Mathematics. Providence, Rhode Island: American Mathematical Society, 2014.
[221] Alfred Tarski. *Einführung in die mathematische Logik und die Methodologie der Mathematik*. Wien: Julius Springer, 1937. URL: https://doi.org/10.1007/978-3-7091-5928-6.
[222] Grzegorz Tomkowicz and Stan Wagon. *The Banach-Tarski Paradox*. Vol. 163. Encyclopedia of Mathematics and Its Applications. Cambridge, New York: Cambridge University Press, 2016. URL: https://doi.org/10.1017/CBO9781107337145.
[223] Rebecca Vinsonhaler. "Teaching calculus with infinitesimals". In: *Journal of Humanistic Mathematics* 6.1 (2016), pp. 249–276. URL: https://scholarship.claremont.edu/jhm/vol6/iss1/17.
[224] Martin Väth. *Nonstandard Analysis*. Basel, Bosten, Berlin: Birkhäuser Verlag, 2007.
[225] Frank Wattenberg. "Unterricht im Infinitesimalkalkül: Erfahrungen in den USA". In: *MU. Der Mathematikunterricht* 4.83 (1983), pp. 7–36.
[226] Hermann Weyl. "Über die neue Grundlagenkrise der Mathematik". In: *Mathematische Zeitschrift* 10.1–2 (1921), pp. 39–79. URL: https://doi.org/10.1007/BF02102305.
[227] Torsten Wilholt. *Zahl und Wirklichkeit: eine philosophische Untersuchung über die Anwendbarkeit der Mathematik*. Paderborn: mentis, 2004.
[228] Timothy Williamson. "How probable is an infinite sequence of heads?" In: *Analysis* 67.3 (2007), pp. 173–180. URL: https://doi.org/10.1093/analys/67.3.173.
[229] W. Hugh Woodin. "In search of ultimate-L the 19th Midrasha Mathematicae Lectures". In: *Bulletin of Symbolic Logic* 23.1 (2017), pp. 1–109. URL: https://doi.org/10.1017/bsl.2016.34.

Index

ACA_0 143, 144
accumulation point 44
actual infinity 146, 147
actually infinite universe 138, 139
Adjunction of Ω 104
aleph operation 148
almost everywhere 67, 74
Archimedean
– axiom 31, 154, 155
– field 155
Arithmetic Comprehension Axiom 144
assignable quantity 15
atom 76
ATR_0 143
axiom
– of boundedness 93
– of choice 23, 165–167
– of constructibility 159, 165, 183
– of countable choice 166
– of dependent choice 166
– of existence 19
– of extensionality 19, 95
– of infinity 22, 95, 146
– of pairing 21, 95
– of power set 21
– of regularity 22
– of relativization 97
– of union 21, 95
axiom schema 21
– of bounded idealization 93
– of collection 95
– of idealization 90
– of relative transfer 98
– of replacement 22
– of separation 19, 95
– of standardization 57, 91
– of transfer 17, 32, 89, 95

background set theory 160, 161
Banach–Tarski paradox 166, 186
bijective 24
Bolzano–Weierstrass theorem 144
Bolzano's theorem 50
Boolean prime ideal theorem 62, 167
Borel
– measure 188
– σ-algebra 188

bounded
– infinite line 150
– sequence 102
– set theory 93, 94
Brownian motion 189
BST 93

canonical extension 180
Cantor–Dedekind postulate 153, 154
cardinal
– addition 149
– exponentiation 149
– multiplication 149
– number 148
cardinality 149
– of a finite set 31
Cartesian product 23
Cauchy sequence 42, 49
chain rule 52
closure principle 112
cluster point 42
cognitive existence 116
compactness theorem 75, 157
completeness theorem, Gödel's 144, 157
comprehensions schema (in Z_2) 143
concurrent 86
conservative
– over PRA 145
– over ZF 23, 61, 62, 167, 184
– over ZFC 88, 94, 178, 184
constant 18
constructible hierarchy 158, 159
constructive analysis 176
constructivism 139, 140
context 98, 111
continuity
– in constructive analysis 176
– of a real function 71, 102
– of a standard function 43
– of an internal function 84, 85
continuum 152–156
continuum hypothesis 134, 158
– generalized 158, 159
convention about contexts 99
convergent 41, 45, 72, 84
correctness theorem 144
countable idealization 61

countably saturated 181
cumulative hierarchy 157, 158

Dedekind-infinite 151
definable models 181
definition principle 112
delta function 83
δ-incomplete filter 87
Δ_1^0-comprehension schema 143
derivative
– of a real function 71
– of a standard function 46
– of an internal function 84, 85
differentiable 46, 71, 84
differentiation rules 51
division points (of a Riemann partition) 47
domain
– of a function 24
– of discourse 17

effective 165
elementary embedding 78
empty set 20
enlargement 86
$\in$-minimal element 22
equipotent 24
event 189
exhaustion method 16
existence principle 112
extended real number system 150
extension principle 107
– elementary 124
external
– element 81
– formula 88
– function 81
– picture 164
– set 81
– statement 19

family of sets 20
field axioms 25
filter 73
finite 102
– cardinal 148
– intersection property 86
– methods 141
– omega number 69
– ordinal 148

– partition 47
– probability space 189
– set 10, 11, 30, 97
finitely satisfiable 86
finitism 139, 140, 146
finitistically reducible 144
forcing method 159
formalism 140–142
formula
– external 88
– internal 81, 88, 100
– transitively bounded 77
Fréchet filter 73
free filter 74
function 24
– continuous 71
– convergent 45
– differentiable 46, 71
– external 81
– integrable 47
– internal 81
fundamental theorem of calculus 55

generalized equality 16
geometric
– series 1
– straight line 154

Hahn–Banach limit 127
Hilbert's program 134, 141
– partial realization of 144, 145
Hrbaček set theory 95–97
HST 95
hyperfinite
– product 71, 83
– sequence 70
– set 70, 82, 97
– sum 71, 83
hyperreal number 74

𝕀 95
Idealization axiom for real numbers 34
illegal
– set formation 21, 38, 88
– transfer 38, 90
inassignable quantity 15, 155
incomparable 154
incompleteness theorems, Gödel's 134, 158
indispensability argument 137

indivisible 63
induction
– in the metalanguage 27
– mathematical 28
induction axiom (in Z_2) 143
induction schema
– in PRA 140
– of Peano arithmetic 162
inductive set 26
infinite
– cardinal 148
– omaga number 69
– ordinal 148
– quantity 15
– set 30
infinitely
– close 34, 70
– fine partition 47
– repeated coin toss 190
– small 1, 15, 34, 69, 102
infinitesimal 1, 15, 34, 63, 69
infinitism 147
injective 24
inner model 159
integrable 47, 84
integral
– of a real function 72, 109
– of a standard function 47
– of an internal function 84, 85
intermediate points (of a Riemann partition) 47
intermediate value theorem 50, 144
internal
– concept 100
– definition principle 81
– element 81
– formula 81, 88, 100
– function 81
– picture 163
– set 81, 95
– set theory 87–93
– statement 19
– statement (in RBST) 111
interpretation 160
intersection
– of a family of sets 20
– of two sets 20

κ-saturated 86, 181

L_1 143
L_2 142
L_2-induction schema 143
law of large numbers 190
least upper bound 37
least-upper-bound property 37, 82
Leibniz's principle 104
limit
– of a real sequence 72
– of a standard function 45
– of a standard sequence 41
– of an internal sequence 84, 85
limit ordinal 148
limited 34
linea infinita
– interminata 150
– terminata 150

mapping 24
– bijective 24
– injective 24
– surjective 24
mathematical induction 28
maximum 11, 31, 50
mesh (of a partition) 47
metalanguage 160
metamathematics 139, 141
metanumber 26, 169, 170
minimum 11, 31, 50
model 160
– inner 159
– special 181
modulus of continuity 176
more geometrico 133
multiverse 164, 165

natural numbers 7–9, 25, 26, 97, 168–172
non-Euclidean geometry 134
nonprincipal filter 74
nonstandard
– embedding 80
– function 83
– number 173–191
– world 81
number
– line 153, 154
– of elements 31
– quantifier 143
– variable 142

object
- language 160
- number 26, 169, 170
- set theory 161
objectivism 147
observable neighbor 98
- principle 112
ω 147
omega number 67
- finite 69
- infinite 69
- infinitely close 70
- infinitely small 69
- infinitesimal 69
- negative infinite 69
- positive infinite 69
ω-model 143
operation 18
operator symbol 18
order axioms 25
ordered pair 23
ordinal number 148

PA 162
pair 21
parameter 18
part-whole axiom 147, 149, 150
Peano
- arithmetic 143, 162
- axioms 168, 169
- structure 168, 169
Π_1^0-formula 143
Π_2^0-sentence 144
Π_1^1-CA$_0$ 143
Platonism 133, 136
polysaturated 87
positive infinite 69
potential infinity 146, 147
power set 21
PRA 140
predicate, n-ary 18
predicativism 146
primitive recursive arithmetic 140
probability 189
probability space 188

quantity
- assignable 15
- infinite 15
- infinitely small 15
- infinitesimal 15
- unassignable 15, 155
quasiorder 97

$\mathbb{R}$ 20
random
- experiment 188, 190
- variable 188, 189
range 24
RBST 97
RCA$_0$ 143
real number 6, 16
- as a primitive notion 17
- in constructive analysis 176
- infinitely close 34
- infinitely small 34
- infinitesimal 34
- limited 34
- unlimited 34
realism
- empirically grounded 137, 138
- in ontology 138
- in truth-value 138
- metaphysical 133, 136, 137
- partial 171
- set-theoretical 136
rectangle property of the integral 55
recursion theorem 10
- for functions 22, 29
- for operations 22
recursive comprehension axiom 143
recursive definitions 30
- in the metalanguage 27
regular sequence 176
regularity over $\mathbb{I}$ (axiom) 95
relation 24
relative
- bounded set theory 97–100
- concepts 98
- observability principle 111
reverse mathematics 142–145
Riemann
- partition 47
- sum 47

$ 95
sample space 188, 189

sequence
- convergent 41, 72
- hyperfinite 70
- regular 176
set 16
- as a primitive notion 17
- difference 20
- empty 20
- external 81
- finite 10, 11, 30, 97
- hyperfinite 70, 82, 97
- infinite 30
- internal 81, 95
- quantifier 143
- theory 156–167
- variable 142
- well-founded 95
σ-additivity 188
Σ_1^0-formula 143
Σ_1^0-induction schema 143
singleton 21
Skolem's paradox 161
smooth infinitesimal analysis 66
Solovay model 182
sorites paradox 40
stability principle 112
standard 16, 17, 88, 97
- core 94
- core interpretability 94
- definition principle 80
- element 80
- function 80
- model 161–163
- parameter 32, 89
- part 36, 71, 108
- part axiom 36
- part principle 108, 110
- picture 164
- set 80, 95
- world 78
standardization 95
*-finite set 82
statement
- closed 18
- external 19
- internal 19
- open 18
stochastic process 189

subset 20
- proper 20
successor ordinal 148
superreal numbers 105–107
superstructure 77
supremum 37
surjective 24
surreal numbers 155
syncategorematic 155

Tarski's ultrafilter theorem 62, 74
transfer principle 107, 110, 174
- for elementary embeddings 79
- for hyperreal numbers 75
transitively bounded formula 77
transitivity of $\mathbb{I}$ (axiom) 95

ultraclose 98, 111
ultrafilter 74
ultrafilter lemma see Tarski's ultrafilter theorem
ultrafinitism 140, 146
ultralarge 98, 111
ultrasmall 98, 111
unbounded infinite line 150
uniform continuity 44, 102
union
- of a family of sets 21
- of two sets 22
universals ante rem 138, 153, 171
universe of discourse 17
unlimited 1, 34
upper bound 37
urelement 76

variables 18
von Neumann's hierarchy see cumulative hierarchy

Weak König's Lemma 144
well-founded set 95
well-order relation 148
well-ordering
- principle 29, 82
- theorem 166
$\mathbb{WF}$ 95
WKL_0 143, 144

Z_2 142
Zorn's lemma 166